AF676127

# Plant Cell Wall Plasticity under Stress Situations

# Plant Cell Wall Plasticity under Stress Situations

Editors

**Penélope García-Angulo**
**Asier Largo-Gosens**

MDPI • Basel • Beijing • Wuhan • Barcelona • Belgrade • Manchester • Tokyo • Cluj • Tianjin

*Editors*
Penélope García-Angulo
Universidad de León
Spain

Asier Largo-Gosens
Universidad Andrés Bello
Chile

*Editorial Office*
MDPI
St. Alban-Anlage 66
4052 Basel, Switzerland

This is a reprint of articles from the Special Issue published online in the open access journal *Plants* (ISSN 2223-7747) (available at: https://www.mdpi.com/journal/plants/special_issues/cell_wall_plasticity).

For citation purposes, cite each article independently as indicated on the article page online and as indicated below:

LastName, A.A.; LastName, B.B.; LastName, C.C. Article Title. *Journal Name* **Year**, *Volume Number*, Page Range.

**ISBN 978-3-0365-5757-1 (Hbk)**
**ISBN 978-3-0365-5758-8 (PDF)**

Cover image courtesy of Asier Largo-Gosens

© 2022 by the authors. Articles in this book are Open Access and distributed under the Creative Commons Attribution (CC BY) license, which allows users to download, copy and build upon published articles, as long as the author and publisher are properly credited, which ensures maximum dissemination and a wider impact of our publications.
The book as a whole is distributed by MDPI under the terms and conditions of the Creative Commons license CC BY-NC-ND.

# Contents

# About the Editors

**Penélope García-Angulo**

Penélope García-Angulo (Plant Physiology Associate Professor, Doctor) obtained a Ph.D. in Biological Sciences from the ULE in 2005 with a cum laude score. From April 2006 to date, she has been carrying out her work as a lecturer and researcher in the plant physiology area of the ULE, first as an assistant professor (from 2006 to 2009), later as a lecturer (from 2009 to 2019), and currently as an associate professor. During the 2004–2020 period, she published 32 scientific papers in journals in the plant science field according to the Journal Citation Reports. She is a co-author of two non-indexed publications with an index of relative quality, two book chapters, one book, and two patents. She has presented 21 communications at international scientific congresses and 22 at national congresses. Moreover, she has participated as a collaborator in nine research projects and three company agreements via article 83. She was (2016–2019) the principal researcher and the coordinator of a challenge-collaboration project with an enterprise. She has been the supervisor of 4 Doctoral theses and 2 Bachelor theses, as well as the tutor of 17 research projects. From 2010 to 2016, she was the Coordinator of the Biotechnology Degree at the University of León, and currently, she is the Coordinator of the Bachelor of Excellence in Sciences at the ULE. For many years, Penélope García-Angulo's research interest has been focused on plant cell wall polymers, such as cellulose, hemicelluloses, pectins, and lignin, and their plasticity under stress situations, as some of the cell wall modifications that take place during stress situations could have important biotechnological applications. As part of this research, she has worked with in vitro cultures of different species, and over the last 5 years, Penélope has been the scientific advisor-in-chief of the Agrovet Healthy Plants Biotechnological Institute, which is a private enterprise focused on the production of certified pathogen-free plants. As part of the certification program, the enterprise has developed, under her supervision, an apical shoot meristem regeneration culture program, different pathogen detection techniques, and chemical fingerprinting protocols. Over the last 5 years, her research has focused on the role of the cell wall in plant defense.

**Asier Largo-Gosens**

Asier Largo-Gosens (Plant Physiology Lecturer, Doctor) is a plant physiology researcher currently specializing in plant molecular biology and biotechnology. He obtained a Ph.D. degree in Biological Sciences at the Universidad de Léon (Spain) in 2016 with a cum laude score (the highest score) and was awarded an Extraordinary Doctorate Award in the sciences by the Universidad de León in 2020. He carried out a short stay at CRAG in Cerdanyola del Vallès (Barcelona, Spain) in 2012. He also participated in a long Postdoctoral stay at the Universidad Andrés Bello (Santiago, Chile) from 2017 to 2021. During his stay, he obtained Postdoctoral FONDECYT project (Conicyt, Chile) funding (approx. 90,000 EUR, 2017–2020). He also worked on two Postdoctoral contracts from the Universidad Andrés Bello (2020 and 2021), and currently, he is a Postdoctoral "María Zambrano" fellow at the Universidad de León under the supervision of Dr. Mélida (Universidad de León).

His research has been focused on understanding the synthesis and metabolism of different carbohydrate components of plant cell walls and on the role of the cell wall of microorganisms in their interactions with plants. During his research, he co-authored 12 scientific papers, all published in Q1 JCR journals. He has participated in three Spanish national projects and one local research project and has supervised one FONDECYT Postdoctoral project of CONICYT (Chile). The results of his research have been presented in 12 scientific meetings. He was the supervisor of one Undergraduate thesis of a Biotechnology degree and one Master's thesis of a Master's in Biotechnology and Life, both degrees of the UNAB (Chile), and he is currently supervising two Undergraduate theses and one Ph.D. student. He has participated in the teaching duties of practical classes in Biology and Biotechnology degrees at the Universidad de León 2012–2013 (20 hours), as an associate professor in the Plant and Environmental Biotechnology subject of the Biotechnology degree from the Universidad Andrés Bello (Chile) in 2021 (45 hours), and in the teaching duties of the Plant Physiology area of the Universidad de León in 2022–2023 (60 hours).

*plants*

*Editorial*

# Plant Cell Wall Plasticity under Stress Situations

**Penélope García-Angulo [1,*] and Asier Largo-Gosens [2,*]**

1 Faculty of Biological and Environmental Sciences, Universidad de León, 24007 Leon, Spain
2 Centro de Biotecnología Vegetal, Universidad Andrés Bello, Santiago 8370186, Chile
* Correspondence: pgara@unileon.es (P.G.-A.); asierlargogosens@gmail.com (A.L.-G.)

This Special Issue, entitled "Plant Cell Wall Plasticity under Stress Situations", is a compilation of five articles, whose authors deepen our understanding of the roles of different cell wall components under biotic and abiotic stress.

The plant cell wall is mainly formed of complex polysaccharides, with multiple interactions between the components that form a network which must be extensible, so as to enable cell expansion, rigid, so as to resist compression and tension forces, and modifiable in response to environmental changes.

Cellulose, the most abundant and resistant polysaccharide on earth, is the main component of the cell wall. The cellulose scaffold is involved in a matrix formed of polysaccharides, such as pectins and hemicelluloses, whose types and proportions vary depending on the species, tissue, and cell type. The deposition of lignin—the second most abundant polymer on earth—in secondary cell walls increases the resistance, leading to growth cessation. All these polymers are crosslinked into the wall in a process that can occur spontaneously and/or by the actions of different modifying enzymes. The control of the synthesis of these cell wall components and/or the interactions between them gives this structure a high plasticity, which is a key factor in the modulation of growth and defense responses under different types of stress.

The evolution of the cell wall allows plants to colonize the land, acquiring the typical erect appearance that allows them to grow taller in order to ensure the absorption of sunlight. Among other important functions, the transport of water and nutrients throughout the plant would not be possible without lignin, which enables the cell wall to be adequately resistant, allowing for long-distance water transport. The deposition and coupling of lignin in the cell wall are controlled by several enzymes, such as laccases and peroxidases. In the first paper of this Special Issue [1], the authors demonstrate the existence of peroxidases with high homology to those of angiosperms in a non-vascular plant, *Physcomitrium patens*. Despite the lack of lignin in this moss, the study describes the discovery of novel class-III peroxidases that could be involved in the biosynthesis of lignin-like polyphenols present in the plant's primitive water transport system. This discovery could be important for the study of the evolution of lignin biosynthesis in vascular plants.

As we have previously mentioned, the cell wall is involved in the control and uptake of water. The hydration capacity of this structure depends on its composition, which eventually alters the water potential of the cell. In the second paper presented in this Special Issue [2], the authors demonstrate, by means of in vitro experiments with bacterial cellulose composites and plant cell wall components, such as pectins and hemicelluloses, that the water potential and degree of hydration depend on the cell wall composition. The authors also demonstrate that the presence of cell wall proteins facilitates wall rehydration and swelling.

Due to the jellification capacity of pectins, researchers have always considered that they play a key role in cell wall rehydration. In the third paper of this Special Issue [3], based on the performance of in vitro and in vivo experiments, the authors demonstrate that pectin cross-linking modifications involving calcium ions and/or boric acid alter

**Citation:** García-Angulo, P.; Largo-Gosens, A. Plant Cell Wall Plasticity under Stress Situations. *Plants* **2022**, *11*, 2752. https://doi.org/10.3390/plants11202752

Received: 30 September 2022
Accepted: 12 October 2022
Published: 18 October 2022

**Publisher's Note:** MDPI stays neutral with regard to jurisdictional claims in published maps and institutional affiliations.

**Copyright:** © 2022 by the authors. Licensee MDPI, Basel, Switzerland. This article is an open access article distributed under the terms and conditions of the Creative Commons Attribution (CC BY) license (https://creativecommons.org/licenses/by/4.0/).

the viscosity of pure pectin standards and prevent their water loss. These modifications affect not only water retention in the cell wall but also its defense capacity against fungal pathogens. In line with this study, the fourth paper [4] demonstrates how the cell wall composition of *Phaseolus vulgaris* L. affects its enzymatic degradation, which ultimately has an effect on the host's susceptibility to *Pseudomonas syringae* pv. *phaseolicola*. Interestingly, cell wall remodeling can be induced artificially by means of priming molecules which increase the resistance of crops to diseases.

Plants modify their cell walls not only during growth and development but also when they are under attack, and several enzymes are involved in this remodeling. In the fifth paper included in this Special Issue [5], the authors review the latest knowledge regarding the roles of plant β-glucanases in the remodeling and turnover of cell walls, as well as in plant defense and adaptative responses.

The cell wall is a resistant but extremely dynamic cell compartment which controls multiple functions, such as defense responses, growth control, water retention, and ion and molecule exchange, among others. Due to its great diversity in terms of its components and interactions, the cell wall has a wide variety of physicochemical and mechanical properties, which means that this structure is very interesting with respect to the development several useful applications. For instance, we wear cell walls, write on cell walls, eat and feed on cell walls, produce heat by burning cell walls, and we can even build housing made of cell walls. With this Special Issue, we encourage the further study of this important and fascinating structure, and we hope that the reader enjoys and learns from reading it.

**Author Contributions:** All authors contributed equally to this article. All authors have read and agreed to the published version of the manuscript.

**Funding:** This research received no external funding.

**Conflicts of Interest:** The authors declare no conflict of interest.

## References

1. Martínez-Cortés, T.; Pomar, F.; Novo-Uzal, E. Evolutionary Implications of a Peroxidase with High Affinity for Cinnamyl Alcohols from *Physcomitrium patens*, a Non-Vascular Plant. *Plants* **2021**, *10*, 1476. [CrossRef] [PubMed]
2. Thompson, D.S.; Islam, A. Plant Cell Wall Hydration and Plant Physiology: An Exploration of the Consequences of Direct Effects of Water Deficit on the Plant Cell Wall. *Plants* **2021**, *10*, 1263. [CrossRef] [PubMed]
3. Forand, A.D.; Finfrock, Y.Z.; Lavier, M.; Stobbs, J.; Qin, L.; Wang, S.; Karunakaran, C.; Wei, Y.; Ghosh, S.; Tanino, K.K. With a Little Help from My Cell Wall: Structural Modifications in Pectin May Play a Role to Overcome Both Dehydration Stress and Fungal Pathogens. *Plants* **2022**, *11*, 385. [CrossRef] [PubMed]
4. De la Rubia, A.G.; Mélida, H.; Centeno, M.L.; Encina, A.; García-Angulo, P. Immune Priming Triggers Cell Wall Remodeling and Increased Resistance to Halo Blight Disease in Common Bean. *Plants* **2021**, *10*, 1514. [CrossRef] [PubMed]
5. Perrot, T.; Pauly, M.; Ramírez, V. Emerging Roles of β-Glucanases in Plant Development and Adaptative Responses. *Plants* **2022**, *11*, 1119. [CrossRef]

plants

MDPI

Article

# Evolutionary Implications of a Peroxidase with High Affinity for Cinnamyl Alcohols from *Physcomitrium patens*, a Non-Vascular Plant

**Teresa Martínez-Cortés [1], Federico Pomar [1] and Esther Novo-Uzal [2,*]**

[1] Grupo de Investigación en Biología Evolutiva, Centro de Investigaciones Científicas Avanzadas, Universidade da Coruña, 15071 A Coruña, Spain; teresa.martinez.cortes@gmail.com (T.M.-C.); federico.pomar@udc.es (F.P.)
[2] Instituto Gulbenkian de Ciência, 2780-156 Oeiras, Portugal
* Correspondence: esther.novo.uzal@gmail.com

**Abstract:** *Physcomitrium (Physcomitrella) patens* is a bryophyte highly tolerant to different stresses, allowing survival when water supply is a limiting factor. This moss lacks a true vascular system, but it has evolved a primitive water-conducting system that contains lignin-like polyphenols. By means of a three-step protocol, including ammonium sulfate precipitation, adsorption chromatography on phenyl Sepharose and cationic exchange chromatography on SP Sepharose, we were able to purify and further characterize a novel class III peroxidase, PpaPrx19, upregulated upon salt and $H_2O_2$ treatments. This peroxidase, of a strongly basic nature, shows surprising homology to angiosperm peroxidases related to lignification, despite the lack of true lignins in *P. patens* cell walls. Moreover, PpaPrx19 shows catalytic and kinetic properties typical of angiosperm peroxidases involved in oxidation of monolignols, being able to efficiently use hydroxycinnamyl alcohols as substrates. Our results pinpoint the presence in *P. patens* of peroxidases that fulfill the requirements to be involved in the last step of lignin biosynthesis, predating the appearance of true lignin.

**Keywords:** *Physcomitrella*; hydroxycinnamyl alcohols; plant evolution; peroxidase; abiotic stress

**Citation:** Martínez-Cortés, T.; Pomar, F.; Novo-Uzal, E. Evolutionary Implications of a Peroxidase with High Affinity for Cinnamyl Alcohols from *Physcomitrium patens*, a Non-Vascular Plant. *Plants* **2021**, *10*, 1476. https://doi.org/10.3390/plants10071476

Academic Editors: Penélope García-Angulo and Asier Largo-Gosens

Received: 20 June 2021
Accepted: 15 July 2021
Published: 19 July 2021

**Publisher's Note:** MDPI stays neutral with regard to jurisdictional claims in published maps and institutional affiliations.

**Copyright:** © 2021 by the authors. Licensee MDPI, Basel, Switzerland. This article is an open access article distributed under the terms and conditions of the Creative Commons Attribution (CC BY) license (https://creativecommons.org/licenses/by/4.0/).

## 1. Introduction

Land colonization by plants and their subsequent diversification was one of the most important events in the history of life. Terrestrialization forced plants to cope with new stresses absent in the aquatic medium, such as UV light and limited water supply. To avoid water loss, plants developed different strategies to accumulate water in their tissues, to supply it or to minimize its loss. These first land plants, such as mosses, were poikilohydric, whose water potential was equilibrated with surrounding water sources [1]. Likewise, the first evolutionary radiation among land plants is related to the diversification of tracheids, which appeared in vascular plants (tracheophytes), about 450 million years ago and they have been defined as single-celled conduits with lignin in their cell wall [2]. Lignin is a polyphenolic polymer that confers structural support and flexural stiffness to the aerial part of the plant and provides water impermeability, including resistance against tensile forces of water columns. Lignins are mainly formed from the oxidative coupling of three *p*-hydroxycinnamyl alcohols: *p*-coumaryl, coniferyl and sinapyl alcohols (monolignols). The cross-coupling reaction of monolignol radicals produces a hydrophobic heteropolymer composed of *p*-hydroxyphenyl (H), guaiacyl (G) and syringyl (S) units [3].

Although lignin has traditionally been linked to vascular plants, polyphenols and lignin-like compounds have been found in species without a true vascular system, such as charophycean green algae [4] and bryophytes [5,6]. Lignin-like compounds are polyphenolic polymers usually detectable with typical methods of lignin determination, such as acetyl bromide or nitrobenzene oxidation, but unlike lignin, they lack β-*O*-4 bonds and aryl-glycerol-β-aryl ether structures. The composition is very variable and in many cases unknown but not related to the three *p*-hydroxycinnamyl alcohols that are considered markers for lignins. This finding implies that at least part of the phenylpropanoid pathway that eventually led to lignin biosynthesis was present in algae, and the presence of lignin in tracheids may only have involved the expression of those genes in a different type of cell [7].

The last step of lignin biosynthesis, the oxidation of monolignols, is driven by laccases [8] and peroxidases [9]. Secretory plant peroxidases (class III Prx) are heme-containing glycoproteins that oxidize diverse substrates using hydrogen peroxide as an electron donor. Peroxidases are usually rich in isoenzymes, generated from post-transcriptional and post-translational modifications [10], with expression patterns usually dependent on development and stress conditions, which make it difficult to assign specific functions to individual peroxidase isoenzymes. Nonetheless, diverse responses to a plethora of stresses or growth conditions have been reported, especially in *Arabidopsis*, indicating specific functions for the different isoforms [9,11].

*Physcomitrium (Physcomitrella) patens* is a bryophyte used as a model organism for evolutionary developmental biology and non-vascular plant studies. *P. patens* shows high tolerance to different environmental cues, such as drought and osmotic and saline stresses, which allows survival in periods when water supply is a limiting factor [12]. RedoxiBase reports 53 class III peroxidases and four pseudogenes in the *P. patens* genome. However, information about *P. patens* peroxidase functions is scarce. The best characterized peroxidase is Prx34 (PpaPrx13 according to RedoxiBase nomenclature), which was reported to play a role upon fungal attack and catalyze ROS production [13,14].

In this paper, we report the purification of one *P. patens* peroxidase, upregulated upon salt and oxidative stresses. This enzyme was further purified and characterized, showing homology to angiosperm peroxidases involved in lignification, and with a catalytic efficiency against coniferyl alcohol, a precursor of lignin, of the same order as angiosperm lignification-related peroxidases, despite the fact that *P. patens* does not contain lignin in its cell walls.

## 2. Results

### 2.1. Abiotic Stress Strongly Modulates the Expression of a Basic Peroxidase in P. patens

*P. patens* is a moss that is highly resistant to abiotic stress, compared to other model plants such as *Arabidopsis thaliana*, and is especially tolerant to desiccation [15], in line with its phylogenetic position as a moss, and whose ancestors were early colonists of land around 500 million years ago. This bryophyte is thus a useful tool to study responses to abiotic stress. Peroxidases are enzymes known to change their expression pattern in response to different types of stress [16,17]. Here, we selected different abiotic stresses and monitored in a time course both peroxidase activity and isoform pattern from protein extracts of *P. patens* gametophores grown in liquid medium. The results (Figure 1A) show that $H_2O_2$ caused an early increase in peroxidase activity, peaking 1h after treatment. The addition of ascorbic acid, a known $H_2O_2$ scavenger, returned peroxidase activity to control levels. NaCl and salicylic acid (SA) also enhanced peroxidase activity 24 h after treatment, going back to control levels afterwards. Mannitol slightly changed peroxidase activity throughout the treatment. Although both osmotic and salt stress can be abscisic acid (ABA) dependent, *P. patens* peroxidase activity in response to ABA treatment did not mirror mannitol or salt stress responses, but strongly decreased from 8 h after the hormone addition. Moreover, SA and mannitol led to the disappearance of a strongly basic peroxidase isoform, which was instead induced by $H_2O_2$ and NaCl (Figure 1B, arrow). We quantified free phenolics as a proxy to measure the stress caused by the different treatments. $H_2O_2$, salt and mannitol treatments significantly enhanced the amount of total phenols in *P. patens* gametophores (Figure 1C). Based on these results, we pursued the purification and further characterization of a strongly basic peroxidase that was induced by $H_2O_2$ and salt, two major stresses faced by the first plants that colonized land.

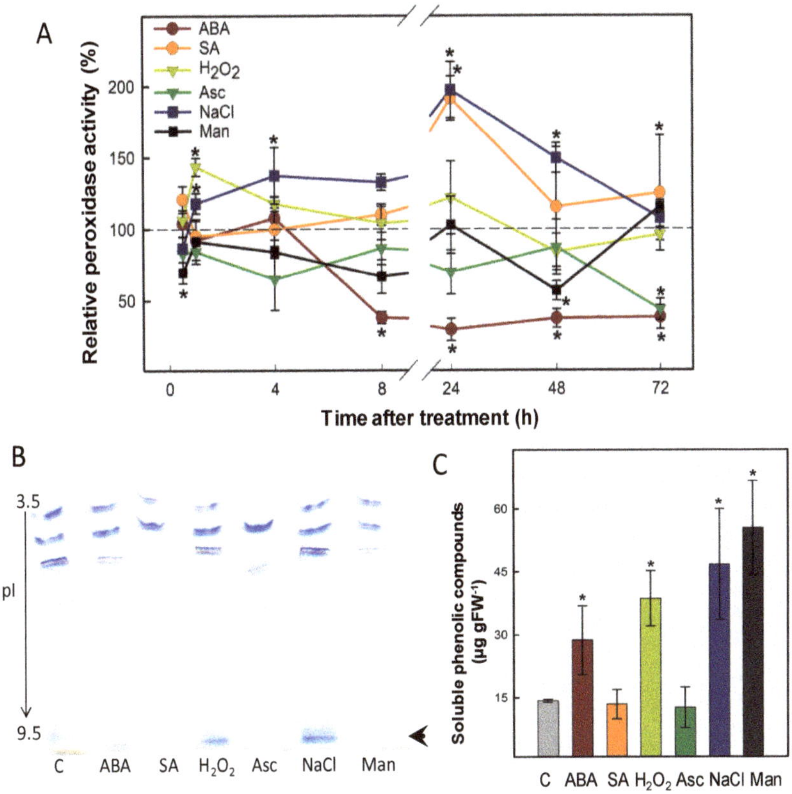


**Figure 1.** Salt and oxidative stress upregulate the expression of a basic peroxidase from *P. patens*. (**A**) Time-course determination of peroxidase activity extracted from *P. patens* gametophores that were subjected to different stress treatments: $H_2O_2$ with or without ascorbic acid (Asc), salt, mannitol (Man), abscisic acid (ABA) and salicylic acid (SA). TMB was used as substrate and peroxidase activity was normalized to control conditions at each time point (shown as a dashed line). (**B**) Effect at 24 h of different treatments on peroxidase isoenzyme pattern revealed by IEF stained with 4-MN. C, control; pI, isoelectric point. (**C**) Effect of stress treatments at 96 h on soluble phenolic content in *P. patens* cultures. Data presented are average values ± SD of $n = 3$ experiments. Statistical analysis was carried out with Duncan's test, asterisks represent statistical differences from control ($p < 0.05$).

## 2.2. *PpaPrx19 Is a 36 kDa Basic Peroxidase*

We extracted total protein from *P. patens* gametophores grown in control conditions and we then followed a three-step protocol for purification, including ammonium sulfate precipitation, hydrophobic chromatography on phenyl Sepharose and cationic chromatography on SP Sepharose. A fractionated precipitation with ammonium sulfate did not allow the partial purification of the peroxidases of interest, which led us to consider just one fraction, precipitating the proteins with 95% $(NH_4)_2SO_4$. The fraction was pooled into phenyl Sepharose chromatography, obtaining two major fractions of peroxidases (Figure 2A). The first eluted fraction (F1) contained only acidic peroxidases, while F2 contained both neutral and basic peroxidases and therefore was selected to continue the purification process (Figure 2). After this step, the specific activity for F2 reached 266 nkat $mg^{-1}$ protein (Table 1). The F2 was then loaded into a cationic exchange chromatography, and the peroxidase bound to SP Sepharose matrix was eluted with a linear gradient of 9.5–11.5 pH. Neutral peroxidases were not retained in the matrix and only one peak of peroxidase activity was eluted, at a pH of 10.9 (Figure 2A). The fraction arising from cationic chromatography migrated as two different bands of 36 and 46 kDa in SDS-PAGE electrophoresis, but an IEF showed only one peroxidase with the pI value determined to be 10.04 (Figure 2B). The peptide mass fingerprinting of the two resultant proteins enabled us to detect that the protein of 46 kDa corresponded to a lipase (accession number XP_001755452). The 36 kDa protein was identified as a predicted protein (access number XP_001781554) which corresponded to PpaPrx19 (according to RedoxiBase nomenclature).

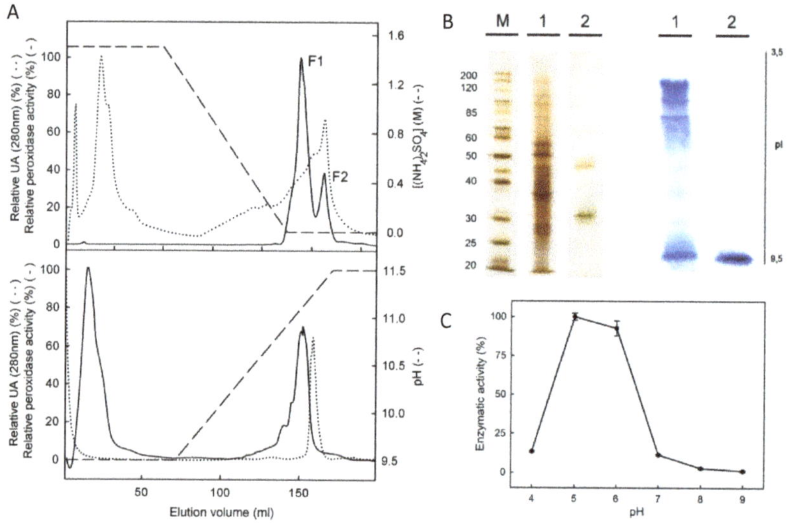


**Figure 2.** Purified PpaPrx19 is a strongly basic peroxidase. (**A**) Purification process of PpaPrx19, including adsorption chromatography on phenyl Sepharose (upper panel, peak F2) and cationic exchange chromatography on SP Sepharose (lower panel). Profiles of peroxidase activity and protein are denoted either by a continuous or by a dotted line, respectively. (**B**) Protein fingerprint in SDS-PAGE (left) and peroxidase isoenzyme pattern in IEF (right) of the crude extract (1) and the purified peroxidase (2) SDS-PAGE and IEF were revealed using the silver staining method and 4-MN in the presence of $H_2O_2$, respectively. (**C**) Dependence on pH of the purified PpaPrx19 activity. Data presented are average values ± SD of $n = 3$ experiments.

**Table 1.** Purification of basic peroxidase PpaPrx19 from *P. patens*. Peroxidase activity was measured using TMB as substrate.

| | Peroxidase Activity (nkat) | Specific Activity (nkat mg$^{-1}$ Protein) | Purification Fold | Yield (%) |
|---|---|---|---|---|
| 95% $(NH_4)_2SO_4$ precipitation | 698 | 28 | 1 | 100 |
| Phenyl Sepharose chromatography | 483 | 266 | 10 | 69 |
| SP Sepharose chromatography | 218 | 23,644 | 850 | 31 |

We confirmed by RT-qPCR that *PpaPrx19* was strongly induced after a treatment with NaCl and that gene expression was modulated in response to other stresses such as hydrogen peroxide and mannitol (Figure S1).

We also evaluated the dependence on pH of PpaPrx19 enzymatic activity, using a different pH (4.0 to 9.0) in the reaction mixture. The purified peroxidase showed the highest activity at pH 5.0, but it rapidly decreased at pH > 6 and showed no activity at pH above 7.0 (Figure 2C). These results do not differ from other peroxidases purified from different sources, with the optimum pH between 4.5 and 6.5 [18–20]. pH is critical for peroxidase activity because pH values outside the optimum prevent the heme from binding to the active site of the enzyme [21].

### 2.3. *PpaPrx19 Is Homologous to Peroxidases with a Role in Lignification*

PpaPrx19 is 332 amino acids long, including a 26 aa N-terminal signal peptide, and it is targeted to the secretory pathway according to analysis with SIGNALP [22] and TARGETP [23] programs. The exon–intron pattern of PpaPrx19 is the second most abundant for *P. patens* and classic for class III peroxidases, consisting in three exons and two introns [13]. In a BLAST search, PpaPrx19 showed the highest identity at the protein level with two other *P. patens* peroxidases (PpaPrx18 and PpaPrx09) and the moss *Tortula ruralis* (Table 2). The rest of the listed peroxidases belong to gymnosperms and angiosperms, and show identity values below 50%, emphasizing the evolutionary distance among them and pointing out the unique characteristics of this peroxidase, at the amino acid level.

**Table 2.** Comparison of PpaPrx19 mature protein sequence with other class III peroxidases. The most similar peroxidase sequences based on BLASTP searches against the RedoxiBase database were used.

| Prx Name | Species | Taxonomical Group | Identity (%) | E-Value |
|---|---|---|---|---|
| PpaPrx18 | *Physcomitrium patens* | Bryophyte | 66 | $3 \times 10^{-144}$ |
| TruPrx01 | *Tortula ruralis* (star moss) | Bryophyte | 65 | $3 \times 10^{-133}$ |
| PpaPrx09 | *Physcomitrium patens* | Bryophyte | 61 | $1 \times 10^{-124}$ |
| CppPrx02 | *Citrus x Paradisi x Poncirus* | Angiosperm | 49 | $3 \times 10^{-97}$ |
| PtaPrx102 | *Pinus taeda* (loblolly pine) | Gymnosperm | 50 | $8 \times 10^{-97}$ |
| CsPrx62 | *Citrus sinensis* | Angiosperm | 49 | $2 \times 10^{-96}$ |
| PabPrx05 | *Picea abies* (Norway spruce) | Gymnosperm | 49 | $6 \times 10^{-95}$ |
| TsPrx15 | *Thellungiella salsuginea* | Angiosperm | 47 | $1 \times 10^{-94}$ |
| BrPrx15-1 | *Brassica rapa* | Angiosperm | 48 | $6 \times 10^{-94}$ |
| PtaPrx28 | *Pinus taeda* (loblolly pine) | Gymnosperm | 46 | $9 \times 10^{-94}$ |
| PabPrx125 | *Picea abies* (Norway spruce) | Gymnosperm | 48 | $4 \times 10^{-93}$ |
| GbPrx04 | *Ginkgo biloba* | Gymnosperm | 49 | $6 \times 10^{-93}$ |

Given these low identity values, we blasted PpaPrx19 against *Arabidopsis* peroxidases, in order to infer a putative function. Most of the peroxidases with the highest identity level have a reported role in lignification (Table 3). This was a surprising result, given that *P. patens* has an internal water-conducting system constituted by hydroids and living cells with thick walls [24] containing pre-lignin and lignin-like polyphenols but no true lignin (defined as the polymerized compounds found in vascular plants) has been described [5,6].

**Table 3.** Reported function of *Arabidopsis* peroxidases which show highest homology to mature PpaPrx19 protein sequence after a BLAST search with Redoxibase.

| Peroxidase Name | TAIR Gene ID | Identity (%) | E-Value | Function | Reference |
|---|---|---|---|---|---|
| AtPrx15 | At2g18150 | 47 | $4 \times 10^{-93}$ | Lignification/abiotic stress | [25] |
| AtPrx49 | At4g36430 | 47 | $6 \times 10^{-93}$ | Lignification | [26] |
| AtPrx53 | At5g06720 | 46 | $1 \times 10^{-92}$ | Lignification | [27] |
| AtPrx14 | At2g18140 | 48 | $2 \times 10^{-92}$ | Biotic stress | [28] |
| AtPrx22 | At2g38380 | 47 | $5 \times 10^{-91}$ | Cold tolerance | [29] |
| AtPrx52 | At5g05340 | 48 | $1 \times 10^{-89}$ | Lignification | [30] |
| AtPrx32 | At3g32980 | 45 | $5 \times 10^{-88}$ | Cell elongation | [31] |
| AtPrx34 | At3g49120 | 45 | $7 \times 10^{-88}$ | Oxidative burst | [32] |

### 2.4. *PpaPrx19 Shows High Affinity for Cinnamyl Alcohols*

The homology that PpaPrx19 shows with peroxidases reportedly involved in lignification, together with the reported presence of lignin-like polyphenols in *P. patens*, led us to characterize this peroxidase based on its preferred substrates, using different well-known peroxidase substrates, including natural precursors of lignin monomers (Table 4). Ascorbic acid is a typical substrate for class I (ascorbate) peroxidases but is poorly oxidized by class III secretory peroxidases [33]. The oxidation of NADH by peroxidases has been associated with cell wall loosening [34]. IAA is an in vitro peroxidase substrate and it has been reported to be catalyzed in vivo in relation to cell growth [35]. Hydroxycinnamic acids such as ferulic acid can be incorporated into suberin [36] and ferulate can also lead to cross-linking of the cell wall [37]. Coniferyl and sinapyl alcohols are polymerized by apoplastic peroxidases to form lignin [38]. Results showed that PpaPrx19 is able to use each assayed substrate, except ascorbic acid and NADH, although IAA was a poor substrate for PpaPrx19. This peroxidase is able to oxidize ferulic acid ($0.53 \pm 0.07$ nkat $\mu g^{-1}$ protein) and sinapyl alcohol ($0.08 \pm 0.01$ nkat $\mu g^{-1}$ protein) but the highest activity is shown using coniferyl alcohol as a substrate ($1.46 \pm 0.13$ nkat $\mu g^{-1}$ protein).

**Table 4.** Enzymatic activities (nkat $\mu g^{-1}$ protein) of purified peroxidase PpaPrx19 in the presence of different substrates. Data presented are average values ± SD of $n$ = 3 experiments. n.d. not detected.

| Substrate | Peroxidase Activity |
|---|---|
| Ascorbic acid | n.d. |
| NADH | n.d. |
| Indole-3-acetic acid (IAA) | $0.01 \pm 0.00$ |
| Ferulic acid | $0.53 \pm 0.07$ |
| Coniferyl alcohol | $1.46 \pm 0.13$ |
| Sinapyl alcohol | $0.08 \pm 0.01$ |

Trying to decipher a putative role in cinnamyl alcohol oxidation, we determined the catalytic parameters of PpaPrx19 for coniferyl and sinapyl alcohols. To calculate the kinetic constants, hydrogen peroxide was used at saturation levels (0.5 mM). The $K_M$ values were calculated according to Lineweaver–Burk equations. For PpaPrx19, apparent $K_M$ values were similar for both alcohols (16.7 µM for coniferyl alcohol and 20.8 µM for sinapyl alcohol). However, $K_{cat}$ is much higher for coniferyl alcohol, rendering a higher catalytic efficiency ($K_{cat}/K_M$), making coniferyl alcohol the best substrate (Table 5).

**Table 5.** Apparent $K_M$, $K_{cat}$ and $K_{cat}/K_M$ values for coniferyl alcohol (CA) and sinapyl alcohol (SA) shown by PpaPrx19.

| Substrate | $K_M$ (µM) | $K_{cat}$ ($s^{-1}$) | $K_{cat}/K_M$ ($\mu M^{-1}$ $s^{-1}$) |
|---|---|---|---|
| CA | 16.7 | 3940.0 | 235.8 |
| SA | 20.8 | 281.0 | 13.5 |

With these extraordinary biochemical characteristics in mind, we searched for structural determinants that define a particular type of isoenzyme, the syringyl peroxidases [39]. We aligned PpaPrx19 with peroxidases with experimental capacity for oxidizing sinapyl alcohol, including ZePrx (the paradigmatic syringyl peroxidase); ATP A2 and HRP which are unable to oxidize sinapyl alcohol; and the three peroxidases that show the highest identity to PpaPrx19, as shown in Table 2. PpaPrx19 not only contains conserved residues important for catalytic mechanisms and the amino acids required for coordination of two $Ca^{2+}$ ions (Figure 3), but it also presents most of the structural determinants of syringyl peroxidases (marked in red in Figure 3), which suggests that this peroxidase has no structural restrictions to oxidizing sinapyl alcohol [39]. As a matter of fact, the PpaPrx19 catalytic properties suggest this peroxidase shows a low $K_M$ for sinapyl alcohol and its ability to oxidize this substrate in vitro (Table 5).

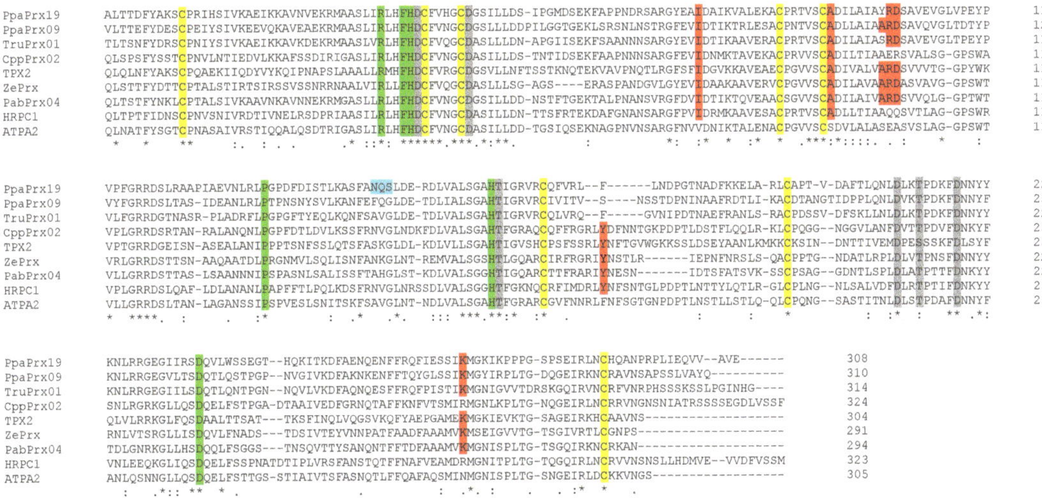

**Figure 3.** PpaPrx19 shows most of residues characteristic of syringyl peroxidases. Amino acid alignment of mature peroxidase sequences, including those purified in this study (PpaPrx19), *Arabidopsis thaliana* ATP A2 (CAA68212), horseradish peroxidase HRPC1 (AAA33377), *Picea abies* PabPrx04 (CAH10839), *Zinnia elegans* ZePrx (CAI54302), *Solanum lycopersicum* TPX2 (AAA65636) and the three peroxidases that show the highest identity to PpaPrx19 (PpaPrx09, TruPrx01 and CppPrx02). Conserved residues important for catalytic mechanisms are shaded in green, the calcium-binding sites are shaded in gray, the S-S bridge-forming cysteines are shaded in yellow, putative *N*-glycosylation site of PpaPrx19 is shaded in blue and structural determinants of syringyl peroxidases are shaded in red. Consensus symbols: '*' indicates fully conserved residues, ':' indicates conserved substitutions and '.' indicates semiconserved residue substitutions.

### 2.5. PpaPrx19 Associates with Lignin Biosynthesis Enzymes and Cell Wall-Related Proteins

Finally, we searched for protein associations by means of STRING (string-db.org). This program provides a network of predicted associations for a particular group of proteins based on high-throughput experimental data, literature and database mining [40]. In the case of PpaPrx19, it is located in the center of a network comprising 10 proteins (Figure S2), which are listed as having unknown functions in the poorly annotated *P. patens* genome. We then searched, for each predicted *P. patens* protein, its closest homolog in *Arabidopsis*. The results are shown in Table 6. The proteins with the highest score are three cinnamyl alcohol dehydrogenases (CADs) and one O-methyl-transferase (OMT). CAD participates in the lignin biosynthetic pathway, catalyzing the conversion of cinnamyl aldehydes into their corresponding alcohols. Unfortunately, the CAD proteins identified by STRING have not been characterized biochemically, although they are known to be expressed in lignifying tissues [41]. The associated OMT has been reported to have high affinity (in the μM range) for a plethora of phenylpropanoids, such as coniferyl alcohol and aldehyde, as well as quercetin [42].

We also performed an analysis in Phytozome (https://phytozome.jgi.doe.gov/pz/portal.html; accessed on 8 July 2021) and searched for coexpression patterns with PpaPrx19. The list (Table S1) comprises phenylalanine ammonia-lyase (PAL), the first enzyme of phenylpropanoid metabolism, which includes the branch that leads to lignin formation [3]. Other enzyme-encoding genes are also coexpressed with PpaPrx19, such as β-1,3-glucanase-related and exostosin heparin sulfate glycosyltransferase-related, both associated with remodeling of the cell wall [43]. Moreover, the WRKY transcription factors have been reported to be involved in the regulation of lignin deposition [44]. These associations support the putative involvement of PpaPrx19 in the formation of lignin or lignin-like compounds.

**Table 6.** List of proteins PpaPrx19 (PP1S306_37V6.1) has interactions with, based on STRING.

| *P. patens* Protein ID | *A. thaliana* Homolog | | Function | Score |
|---|---|---|---|---|
| PP1S84_209V6.1 | CADG | AT1G72680.1 | Cinnamyl alcohol dehydrogenase | 0.650 |
| PP1S126_185V6.1 | CAD9 | AT4G39330.1 | Cinnamyl alcohol dehydrogenase | 0.650 |
| PP1S163_63V6.1 | ELI3-2 | AT4G37990.1 | Cinnamyl alcohol dehydrogenase | 0.650 |
| PP1S56_71V6.1 | OMT1 | AT5G54160.1 | O-methyltransferase | 0.650 |
| PP1S123_38V6.1 | | AT1G78780 | PR protein | 0.546 |
| PP1S34_74V6.1 | | AT1G29850 | DNA binding | 0.546 |
| PP1S141_102V6.3 | MBF1A | AT2G42680.1 | Multiprotein bridging factor | 0.531 |
| CHI | | AT2G43590 | Chitinase | 0.521 |
| PP1S96_94V6.1 | MBF1B | AT3G58680.1 | Multiprotein bridging factor | 0.400 |
| PP1S35_215V6.1 | MBF1B | AT3G58680.1 | Multiprotein bridging factor | 0.400 |

## 3. Discussion

The cell wall is characteristic of all plant cells, although its composition varies depending on the cell type, the lineage and environmental conditions. Therefore, the plant cell wall is essential for cell development and in responses to stress, being able to plastically adapt to the cell's needs. Several innovations that arose during plant evolution, such as lignin and suberin, help to promote this plasticity. Lignin is thought to have emerged with vascular plants 450 million years ago, but lignin-like or pre-lignin compounds have been detected in bryophytes and algae [5,45]. The last step of lignin biosynthesis, the oxidation of hydroxycinnamyl alcohols, is catalyzed by class III peroxidases. In this work, we report the purification of a *P. patens* peroxidase, PpaPx19, with the ability to oxidize hydroxycinnamyl alcohols (Tables 4 and 5). This characteristic is surprising not just because *P. patens* does not lignify, but also because of the atypical kinetic properties shown by this peroxidase. In our experiments, PpaPrx19 showed $K_M$ values for hydroxycinnamyl alcohols similar to other peroxidases involved in lignification, from flowering and non-flowering plants. While PpaPrx19 showed $K_M$ values of 16.7 and 20.8 µM for coniferyl and sinapyl alcohols, respectively (Table 5), the gymnosperm *Picea abies* contains two basic peroxidases involved in lignification with reported $K_M$ values for coniferyl alcohol of 16.7 and 23.2 µM [46]. In angiosperms, $K_M$ values have been reported for zinnia (83 µM for coniferyl alcohol) and tomato (11.4 µM for syringaldazine, a chemical analog of sinapyl alcohol) [47,48].

The use of catalytic efficiency ($K_{cat}/K_M$) is preferable in order to compare diverse enzymes and substrates, although very few peroxidase reports calculate this parameter to evaluate enzyme kinetics. In the lycophyte *Selaginella*, two basic peroxidases show values of 3.55 and 28.63 $\mu M^{-1}s^{-1}$ with coniferyl alcohol [20]. GbPrx09 from *Ginkgo biloba*, a gymnosperm, showed values of 4.91 $\mu M^{-1}s^{-1}$ for coniferyl alcohol [17]. In dicots, ZePrx from *Z. elegans* showed a $K_{cat}/K_M$ ratio for coniferyl alcohol of 1.20 $\mu M^{-1}s^{-1}$ [48] and TPX1 (from tomato) showed a $K_{cat}/K_M$ for syringaldazine of 1.50 $\mu M^{-1}s^{-1}$ [47]. In monocots, PviPRX9 from *Panicum virgatum* showed a $K_{cat}/K_M$ ratio for coniferyl alcohol of 1.60 $\mu M^{-1}s^{-1}$ [49].

The $K_{cat}/K_M$ value obtained for PpaPrx19 using coniferyl alcohol as a substrate is not only higher than for peroxidases involved in lignification [48], but also for other enzymes involved in phenylpropanoid metabolism, such as PAL (the first enzyme of the route, [50]), CCR (the first committed enzyme of lignin biosynthesis, [51]) and CAld5H, involved in last steps of lignin biosynthesis [52].

Moreover, our results indicate that PpaPrx19 is able to efficiently use sinapyl alcohol as a substrate. While coniferyl alcohol is easily oxidized by most peroxidases, the capacity of these enzymes to oxidize sinapyl alcohol is not such a common fact and defined a new subgroup named syringyl peroxidases [39]. PpaPrx19 has most of the structural determinants of this new subgroup [39] in the protein primary structure (Figure 3, shaded in red). These structural motifs determine the syringyl oxidase activity shown by peroxidases, but are absent in the two paradigmatic G peroxidases, ATP A2 and HRP A2 from *Arabidopsis* and horseradish, respectively. These structural motifs agree with the experimental capacity of PpaPrx19 of oxidizing sinapyl alcohol in vitro.

All these data strongly suggest that the peroxidase PpaPrx19 may have been involved in lignin biosynthesis, if such a pathway was present in *P. patens*, i.e., PpaPrx19 fulfills the kinetic and structural requirements to oxidize coniferyl alcohol. The presence of an enzyme involved in the biosynthetic route of a compound that appeared later in an evolutionary context is not surprising. Thermospermine emerges as one example of a metabolite typical of vascular plants recently described in non-vascular plants. The only reported function for thermospermine is the regulation of xylem cell maturation, which makes the function it may have in non-vascular plants unclear [53]. It is widely accepted that promiscuous enzymes with several putative substrates are more likely to be recruited to novel metabolic routes [54]. Therefore, a peroxidase with multiple substrates but with particular affinity for coniferyl alcohol would be a good candidate to participate in what eventually would constitute the pathway leading to lignin formation.

Several reports [55–57] indicate that the first appearance of the entire lignin biosynthesis pathway enzymes (excluding the pathway that leads to syringyl lignin formation), from the catalysis of phenylalanine to coniferyl alcohol formation, took place in mosses (*P. patens*), regardless the fact that *P. patens* does not accumulate lignin in the cell wall. Thus, the *P. patens* genome has all the genes necessary for the biosynthesis of lignin, and according to the results presented in this paper, at least some of the enzymes are expressed and functional, but the route does not take place and lignin is not polymerized. Nonetheless, a pre-lignin pathway has been recently suggested, revealing a role of caffeate units for the formation of the *P. patens* cuticle, coupled with ascorbate metabolism [6]. This finding suggests that the biosynthesis of lignin-like or pre-lignin compounds may not originate from the precursors described for canonical lignin and that the enzymes involved in its synthesis have broader specificity than the enzymes participating in true lignin from vascular plants, making lignin evolution an exciting field to explore.

PpaPrx19 may have a function that in vascular plants was later derived for involvement in lignification. This hypothesis is supported by its structural and kinetic homology to peroxidases with an already described role in lignin biosynthesis, such as ZePrx [39,48], and association with other enzymes of the lignin biosynthetic route. The appearance of a primitive water-conducting system, together with stomata and cuticle, were innovations developed by plants during the transition from water to land. Renault [6] already proved the existence of a pre-lignin pathway involved in the formation of the *P. patens* cuticle. Given all this, although the actual function of PpaPrx19 in *P. patens* physiology remains unclear and should be further studied, it is likely involved in the remodeling of the cell wall in response to environmental stress, based on this peroxidase ability of oxidizing phenolic compounds and its upregulation upon several conditions related to water deficiency, the paradigmatic stress for poikilohydric plants lacking a true vascular system.

## 4. Materials and Methods

### 4.1. Plant Material and Treatments

*P. patens* was provided by the Biotechnology Department of the Universidad Politécnica (Madrid, Spain) and cultured as described previously [58]. *P. patens* gametophores were maintained in solid medium under standard conditions in a growth chamber at 25 °C with a 16 h photoperiod. For stress treatments, *P. patens* gametophores were grown in liquid medium on a rotator shaker (130 rpm) and then transferred into medium supplemented with 5 mM $H_2O_2$ (with or without 10 mM ascorbic acid), 250 mM NaCl, 250 mM mannitol, 10 μM abscisic acid (ABA) and 1 mM salicylic acid (SA), respectively [59–62].

### 4.2. Protein Extraction and Precipitation

Gametophore samples (400 g) were homogenized in 50 mM Tris-HCl buffer (pH 7.5) containing 1 mM EDTA, 1 M KCl, and 0.05 g PVPP per g of tissue. The homogenate was filtered through nylon layers and centrifuged at 27,000× *g* for 30 min at 4 °C. The supernatant was dialyzed on cellulose membranes. After protein precipitation with 95% saturation ammonium sulfate, the precipitate was resuspended in 50 mM Tris-HCl pH 7.5 and dialyzed overnight against the same buffer.

### 4.3. Purification of PpaPrx19

Purification of *P. patens* peroxidase 19 was performed in the AKTA System (GE Healthcare, Barcelona, Spain). The dialyzed sample was concentrated in Amicon® Ultra (Merck Millipore, Barcelona, Spain) and dissolved in 1.5 M $(NH_4)_2SO_4$. In the first step, hydrophobic chromatography, proteins were separated on a phenyl Sepharose™ 6 Fast Flow (GE Healthcare, Barcelona, Spain) 31.5 × 1.0 cm gel bed column at a flow rate 1.0 mL·min$^{-1}$, and fractions of 5.0 mL were recovered. The eluent chromatography program was as follows: from 0 to 160 min (100% A, 0% B), from 160 to 360 min (0% to 100% B) and from 360 to 515 min (100% B), where buffer A was 50 mM Tris-HCl (pH 7.5) containing 1.5 M $(NH_4)_2SO_4$, and buffer B was 50 mM Tris-HCl (pH 7.5).

The second step involved ion-exchange chromatography on SP Sepharose Fast Flow (GE Healthcare). To do so, the peroxidase-enriched fractions obtained from the hydrophobic chromatography were dialyzed against 50 mM CAPS, (pH 9.5), and loaded on a 17.5 × 1.0 cm gel bed column equilibrated with 50 mM CAPS (pH 9.5), at a flow rate of 1.0 mL min$^{-1}$. Fractions of 1.0 mL were recovered. The eluent chromatography program was as follows: from 0 to 70 min (100% A, 0% B), from 70 to 170 min (0–100% B) and from 170 to 330 min (100% B), where buffer A was 50 mM CAPS (pH 9.5), and buffer B was 50 mM CAPS (pH 11.5).

### 4.4. Peroxidase Activity Determination

Peroxidase activity was determined in a spectrophotometer at 25 °C in a reaction medium containing 50 mM sodium acetate buffer (pH 5.0) and 0.5 mM $H_2O_2$ with tetramethylbenzidine (TMB) 0.1 mg·mL$^{-1}$ as substrate ($\varepsilon_{652}$ = 39.0 mM$^{-1}$·cm$^{-1}$).

Peroxidase activity was also calculated using different substrates: 100 µM coniferyl alcohol ($\varepsilon_{262}$ = 9.75 mM$^{-1}$·cm$^{-1}$), 100 µM sinapyl alcohol ($\varepsilon_{271}$ = 4.14 mM$^{-1}$·cm$^{-1}$), 50 µM ferulic acid ($\varepsilon_{310}$ = 16.6 mM$^{-1}$·cm$^{-1}$), 200 µM indole-3-acetic acid ($\varepsilon_{261}$ = 3.2 mM$^{-1}$·cm$^{-1}$), 1 mM ascorbic acid ($\varepsilon_{290}$ = 2.8 mM$^{-1}$·cm$^{-1}$) and 150 µM NADH ($\varepsilon_{340}$ = 6.22 mM$^{-1}$·cm$^{-1}$), as described elsewhere [48,63]. For measuring peroxidase activity with ascorbic acid, 50 mM phosphate buffer (pH 7.0) was used.

The pH-dependent enzymatic activity was assessed using 50 mM sodium acetate buffer for pH 4.0 to 6.0 and 50 mM Tris-HCl for pH 7.0 to 9.0, with TMB as substrate.

### 4.5. Kinetic Data Analysis

For determination of apparent $K_M$, 5–100 µM (coniferyl alcohol) and 1–110 µM (sinapyl alcohol) concentrations of each substrate were used in 50 mM sodium acetate buffer pH 5.0. $K_M$ values were determined from the Lineweaver–Burk equation, a linear transformation of the Michaelis–Menten equation.

### 4.6. Electrophoretic Analysis

Isoelectric focusing was performed on a Pharmacia Multiphor II system using Ampholine PAGplate (pH = 3.5–9.5) polyacrylamide gels according to the manufacturer's instructions (GE Healthcare). Peroxidase isoenzymes were stained with 4-methoxy-α-naphthol (4-MN) in the presence of $H_2O_2$.

SDS-PAGE was performed on 10 % (*w*/*v*) polyacrylamide gels using a MiniProtean® 3 Cell electrophoresis kit (Bio-Rad Laboratories, Barcelona, Spain) and a pH 8.8 electrophoresis buffer composed of 192 mM glycine and 25 mM Tris containing 0.1% SDS. SDS-PAGE was performed at 200 V for 40–45 min at room temperature. Proteins were stained with a Plus One Silver Staining kit, according to the manufacturer's instructions (GE Healthcare).

### 4.7. *Phenolic Compound Extraction and Quantification*

Gametophores grown in liquid culture were harvested 96 h after treatment, ground in liquid nitrogen and extracted for soluble phenolic content as previously described [17]. The quantitative determination of phenolics was performed using Folin–Ciocalteu reagent with ferulic acid as standard.

### 4.8. *Molecular Weight MALDI-TOF/TOF*

The purified peroxidase was analyzed in a MALDI-TOF/TOF instrument as previously described [20]. The search for peptide mass fingerprints and tandem MS spectra was performed in the NCBInr database without taxonomy restriction. Mascot scores for all protein identifications were higher than the accepted threshold for significance (at the $p < 0.050$ level, positive rate measured to be 0.047).

### 4.9. *Sequence Data Analysis*

The presence of signal peptides was predicted using TargetP (http://www.cbs.dtu.dk/services/TargetP/, accessed on 24 November 2020), Bacello (http://gpcr2.biocomp.unibo.it/bacello/index.htm, accessed on 24 November 2020) and TargetLoc (http://abi.inf.unituebingen.de/Services/MultiLoc/index_html, accessed on 24 November 2020). pI prediction was carried out using Compute pI/MW tool from ExPASy (http://web.expasy.org/compute_pi, accessed on 24 November 2020). A search for protein sequences homologies was performed using BLASTP from Redoxibase (http://www.peroxibase.toulouse.inra.fr, accessed on 4 February 2021) and protein alignments were carried out using Clustal Omega (https://www.ebi.ac.uk/Tools/msa/clustalo/, accessed on 4 February 2021). Search for *N*-glycosylation sites was performed using the NetNGlyc tool (http://www.cbs.dtu.dk/services/NetNGlyc, accessed on 25 November 2020).

### 4.10. *Gene Expression Analysis by RT-qPCR*

Total RNA was isolated from gametophores grown in liquid medium, 24h after treatment, with Trizol (Invitrogen, Madrid, Spain) essentially as described in [64]. cDNA was synthesized from 200 ng of total RNA using the iScript™ cDNA Synthesis Kit (Bio Rad Laboratories, Barcelona, Spain). For RT-qPCR, the constitutively expressed *18S* gene was used as a reference gene (Fwd 5′-GGACCGATAGGTCTGGGTAA-3′ and Rev 5′-GCAATCCGAAAACTTCACCG-3′) and for *PpaPrx19* amplification, the primers Fwd 5′- CTCACCACTGACTTCTACGC-3′ and Rev 5′-TGGGATGCTGTCCAAGAGTA-3′ were used. The PCR reaction contained Bio-Rad 1x iQ SYBR Green Supermix, 0.3 μM primer mix and 2.5 μg of cDNA for a 50 μL final volume. The PCR program comprised a 1 min denaturation step at 94 °C followed by 40 cycles of amplification (94 °C for 30 s, 58 °C for 45 s and 72 °C for 1 min) and a final elongation step of 6 min at 72 °C. Bio-Rad Optical System Software 3.0 was used for data analysis, and relative expression values were calculated from the resulting Ct values [65].

### 4.11. *Network Construction of PpaPrx19*

The Search Tool for the Retrieval of Interacting Genes/Proteins database (STRING v11) was used to construct the network associated with PpaPrx19. The sequence of PpaPrx19 was loaded in the database and STRING generated a protein–protein interaction (PPI) network with proteins that have interactions with the target protein. Interactions were obtaining with medium confidence from curated databases and textmining.

**Supplementary Materials:** The following are available online at https://www.mdpi.com/article/10.3390/plants10071476/s1, Figure S1: Effect of different abiotic stresses on *PpaPrx19* gene expression, Figure S2: Graphical representation of the proteins PpaPrx9 (PP1S306_37V6.1) has interactions with. Table S1. List of coexpression patterns obtained for PpaPrx19 in Phytozome (https://phytozome.jgi.doe.gov/pz/portal.html), accessed on 8 July 2021.

**Author Contributions:** Conceptualization, F.P., E.N.-U., T.M.-C.; methodology, T.M.-C.; formal analysis, T.M.-C., E.N.-U.; investigation, T.M.-C.; writing—original draft preparation, E.N.-U.; writing—review and editing, T.M.-C., E.N.-U., F.P.; supervision, E.N.-U., F.P.; funding acquisition, E.N.-U., F.P. All authors have read and agreed to the published version of the manuscript.

**Funding:** This research was funded by Xunta de Galicia (Spain), grant number INCITE 08PXIB103182PR. E.N.-U. holds an FCT postdoctoral fellowship (SFRH/BPD/112587/2015).

**Data Availability Statement:** The data is contained within the article.

**Acknowledgments:** *P. patens* was kindly provided by Alonso Rodríguez-Navarro from the Biotechnology Department of Universidad Politécnica (Madrid, Spain). The proteomic analysis assistance of Jesús Mateos (INIBIC, A Coruña, Spain) is gratefully acknowledged.

**Conflicts of Interest:** The authors declare no conflict of interest. The funders had no role in the design of the study; in the collection, analyses, or interpretation of data; in the writing of the manuscript, or in the decision to publish the results.

## References

1. Charron, A.J.; Quatrano, R.S. Between a rock and a dry place: The water-stressed moss. *Mol. Plant* **2009**, *2*, 478–486. [CrossRef]
2. Kenrick, P.R.; Crane, P.R. The origin and early evolution of plants on land. *Nature* **1997**, *389*, 33–39. [CrossRef]
3. Vanholme, R.; De Meester, B.; Ralph, J.; Boerjan, W. Lignin biosynthesis and its integration into metabolism. *Curr. Opin. Biotechnol.* **2019**, *56*, 230–239. [CrossRef] [PubMed]
4. Sørensen, I.; Pettolino, F.A.; Bacic, A.; Ralph, J.; Lu, F.; O'Neill, M.A.; Fei, Z.; Rose, J.K.; Domozych, D.S.; Willats, W.G. The charophycean green algae provide insights into the early origins of plant cell walls. *Plant J.* **2011**, *68*, 201–211. [CrossRef]
5. Espiñeira, J.M.; Novo-Uzal, E.; Gómez-Ros, L.V.; Carrión, J.S.; Merino, F.; Ros Barceló, A.; Pomar, F. Distribution of lignin monomers and the evolution of lignification among lower plants. *Plant Biol.* **2011**, *13*, 59–68. [CrossRef] [PubMed]
6. Renault, H.; Alber, A.; Horst, N.A.; Basilio Lopes, A.; Fich, E.A.; Kriegshauser, L.; Wiedemann, G.; Ullmann, P.; Herrgott, L.; Erhardt, M.; et al. A phenol-enriched cuticle is ancestral to lignin evolution in land plants. *Nat. Commun.* **2017**, *8*, 14713. [CrossRef]
7. Weng, J.K.; Chapple, C. The origin and evolution of lignin biosynthesis. *N. Phytol.* **2010**, *187*, 273–285. [CrossRef]
8. Berthet, S.; Demont-Caulet, N.; Pollet, B.; Bidzinski, P.; Cézard, L.; Le Bris, P.; Borrega, N.; Hervé, J.; Blondet, E.; Balzergue, S.; et al. Disruption of LACCASE4 and 17 results in tissue-specific alterations to lignification of *Arabidopsis thaliana* stems. *Plant Cell* **2011**, *23*, 1124–1137. [CrossRef]
9. Shigeto, J.; Tsutsumi, Y. Diverse functions and reactions of class III peroxidases. *N. Phytol.* **2016**, *209*, 1395–1402. [CrossRef]
10. Welinder, K.G.; Justesen, A.F.; Kjaersgard, I.V.; Jensen, R.B.; Rasmussen, S.K.; Jespersen, H.M.; Duroux, L. Structural diversity and transcription of class III peroxidases from *Arabidopsis thaliana*. *Eur. J. Biochem.* **2002**, *269*, 6063–6081. [CrossRef] [PubMed]
11. Cosio, C.; Dunand, C. Specific functions of individual class III peroxidase genes. *J. Exp. Bot.* **2009**, *60*, 391–408. [CrossRef] [PubMed]
12. Rensing, S.A.; Goffinet, B.; Meyberg, R.; Wu, S.-Z.; Bezanilla, M. The Moss *Physcomitrium (Physcomitrella) patens*: A model organism for non-seed plants. *Plant Cell* **2020**, *32*, 1361–1376. [CrossRef]
13. Lehtonen, M.T.; Akita, M.; Kalkkinen, N.; Ahola-Iivarinen, E.; Rönnholm, G.; Somervuo, P.; Thelander, M.; Valkonen, J.P.T. Quickly-released peroxidase of moss in defense against fungal invaders. *N. Phytol.* **2009**, *183*, 432–443. [CrossRef] [PubMed]
14. Lehtonen, M.T.; Akita, M.; Frank, W.; Reski, R.; Valkonen, J.P. Involvement of a class III peroxidase and the mitochondrial protein TSPO in oxidative burst upon treatment of moss plants with a fungal elicitor. *Mol. Plant Microbe Interact.* **2012**, *25*, 363–371. [CrossRef] [PubMed]
15. Frank, W.; Ratnadewi, D.; Reski, R. *Physcomitrella patens* is highly tolerant against drought, salt and osmotic stress. *Planta* **2005**, *220*, 384–394. [CrossRef]
16. Su, P.; Yan, J.; Li, W.; Wang, L.; Zhao, J.; Ma, X.; Li, A.; Wang, H.; Kong, L. A member of wheat class III peroxidase gene family, TaPRX-2A, enhanced the tolerance of salt stress. *BMC Plant Biol.* **2020**, *20*, 392. [CrossRef]
17. Novo-Uzal, E.; Gutierrez, J.; Martinez-Cortes, T.; Pomar, F. Molecular cloning of two novel peroxidases and their response to salt stress and salicylic acid in the living fossil *Ginkgo biloba*. *Ann. Bot.* **2014**, *114*, 923–936. [CrossRef] [PubMed]
18. Mika, A.; Lüthje, S. Properties of guaiacol peroxidase activities isolated from corn root plasma membranes. *Plant Physiol.* **2003**, *132*, 1489–1498. [CrossRef] [PubMed]
19. Sergio, L.; Cardinali, A.; Di Paola, A.; Di Venere, D. Biochemical properties of soluble and bound peroxidases from artichoke heads and leaves. *Food Technol. Biotechnol.* **2009**, *47*, 32–38.
20. Martinez-Cortés, T.; Pomar, F.; Espiñeira, J.M.; Merino, F.; Novo-Uzal, E. Purification and kinetic characterization of two peroxidases of *Selaginella martensii* Spring. involved in lignification. *Plant Physiol. Biochem.* **2012**, *52*, 130–139. [CrossRef]
21. Adams, J.B. Regeneration and kinetics of peroxidase inactivation. *Food Chem.* **1997**, *60*, 201–206. [CrossRef]
22. Petersen, T.N.; Brunak, S.; Von Hijne, G.; Nielsen, H. SIGNALP 4.0: Discriminating signal peptides from transmembrane regions. *Nat. Methods* **2011**, *8*, 785–786. [CrossRef]
23. Emanuelsson, O.; Nielsen, H.; Brunak, S.; Heijne, G.V. Predicting subcellular localization of proteins based on their N-terminal amino acid sequence. *J. Mol. Biol.* **2000**, *300*, 1005–1016. [CrossRef]

24. Ligrone, R.; Ducket, J.G.; Renzaglia, K.S. Conducting tissues and phyletic relationships of bryophytes. *Philos. Trans. R. Soc. Lond. B Biol. Sci.* **2000**, *355*, 795–813. [CrossRef]
25. Ge, S.; Han, X.; Xu, X.; Shao, Y.; Zhu, Q.; Liu, Y.; Du, J.; Xu, J.; Zhang, S. WRKY15 suppresses tracheary element differentiation upstream of VND7 during xylem formation. *Plant Cell* **2020**, *32*, 2307–2324. [CrossRef]
26. Herrero, J.; Esteban-Carrasco, A.; Zapata, J.M. Looking for *Arabidopsis thaliana* peroxidases involved in lignin biosynthesis. *Plant Physiol. Biochem.* **2013**, *67*, 77–86. [CrossRef] [PubMed]
27. Østergaard, L.; Teilum, K.; Mirza, O.; Mattsson, O.; Petersen, M.; Welinder, K.G.; Mundy, J.; Gajhede, M.; Henriksen, A. Arabidopsis ATP A2 peroxidase. Expression and high-resolution structure of a plant peroxidase with implications for lignification. *Plant Mol. Biol.* **2000**, *44*, 231–243. [CrossRef] [PubMed]
28. Kakrana, A.; Kumar, A.; Satheesh, V.; Abdin, M.Z.; Subramaniam, K.; Bhattacharya, R.C.; Srinivasan, R.; Sirohi, A.; Jain, P.K. Identification, validation and utilization of novel nematode-responsive root-specific promoters in Arabidopsis for inducing host-delivered RNAi mediated root-knot nematode resistance. *Front. Plant Sci.* **2017**, *8*, 2049. [CrossRef] [PubMed]
29. Kim, B.H.; Kim, S.Y.; Nam, K.H. Genes encoding plant-specific class III peroxidases are responsible for increased cold tolerance of the brassinosteroid-insensitive 1 mutant. *Mol. Cells* **2012**, *34*, 539–548. [CrossRef] [PubMed]
30. Fernández-Pérez, F.; Pomar, F.; Pedreño, M.A.; Novo-Uzal, E. The suppression of AtPrx52 affects fibers but not xylem lignification in Arabidopsis by altering the proportion of syringyl units. *Physiol. Plant* **2015**, *154*, 395–406. [CrossRef] [PubMed]
31. Irshad, M.; Canut, H.; Borderies, G.; Pont-Lezica, R.; Jamet, E. A new picture of cell wall protein dynamics in elongating cells of *Arabidopsis thaliana*: Confirmed actors and newcomers. *BMC Plant Biol.* **2008**, *8*, 94. [CrossRef]
32. Bindschedler, L.V.; Whitelegge, J.P.; Millar, D.J.; Bolwell, G.P. A two component chitin-binding protein from French bean: Association of a proline-rich protein with a cysteine-rich polypeptide. *FEBS Lett.* **2006**, *580*, 1541–1546. [CrossRef] [PubMed]
33. De Gara, L. Class III peroxidases and ascorbate metabolism in plants. *Phytochem. Rev.* **2004**, *3*, 195–205. [CrossRef]
34. Liszkay, A.; van der Zalm, E.; Schopfer, P. Production of reactive oxygen intermediates ($O_2^-$), $H_2O_2$, and OH) by maize roots and their role in wall loosening and elongation growth. *Plant Physiol.* **2004**, *36*, 3114–3123. [CrossRef]
35. Gazaryan, I.G.; Lagrimini, L.M.; Ashby, G.A.; Thorneley, N.F. Mechanism of indole-3-acetic acid oxidation by plant peroxidases: Anaerobic stopped-flow spectrophotometric studies on horseradish and tobacco peroxidases. *Biochem. J.* **1996**, *313*, 841–847. [CrossRef]
36. Hiraga, S.; Sasaki, K.; Ito, H.; Ohashi, Y.; Matsui, H. A large family of class III plant peroxidases. *Plant Cell Physiol.* **2001**, *42*, 462–468. [CrossRef]
37. Hatfield, R.D.; Rancour, D.M.; Marita, J.M. Grass cell walls: Story of cross-linking. *Front. Plant Sci.* **2017**, *7*, 2056. [CrossRef] [PubMed]
38. Fagerstedt, K.V.; Kukkola, E.M.; Koistinen, V.V.; Takahashi, J.; Marjamaa, K. Cell wall lignin is polymerised by class III secretable plant peroxidases in Norway spruce. *J. Integr. Plant Biol.* **2010**, *52*, 186–194. [CrossRef] [PubMed]
39. Ros Barceló, A.; Gómez Ros, L.V.; Esteban-Carrasco, A. Looking for syringyl peroxidases. *Trends Plant Sci.* **2007**, *12*, 486–491. [CrossRef] [PubMed]
40. Szklarczyk, D.; Gable, A.L.; Lyon, D.; Junge, A.; Wyder, S.; Huerta-Cepas, J.; Simonovic, M.; Doncheva, N.T.; Morris, J.H.; Bork, P.; et al. STRING v11: Protein-protein association networks with increased coverage, supporting functional discovery in genome-wide experimental datasets. *Nucleic Acids Res.* **2019**, *47*, D607–D613. [CrossRef]
41. Kim, S.J.; Kim, K.W.; Cho, M.H.; Franceschi, V.R.; Davin, L.D.; Lewis, N.G. Expression of cinnamyl alcohol dehydrogenases and their putative homologues during *Arabidopsis thaliana* growth and development: Lessons for database annotations? *Phytochemistry* **2007**, *68*, 1957–1974. [CrossRef] [PubMed]
42. Moinuddin, S.G.; Jourdes, M.; Laskar, D.D.; Ki, C.; Cardenas, C.L.; Kim, K.W.; Zhang, D.; Davin, L.B.; Lewis, N.G. Insights into lignin primary structure and deconstruction from *Arabidopsis thaliana* COMT (caffeic acid O-methyl transferase) mutant *Atomt1*. *Org. Biomol. Chem.* **2010**, *8*, 3928–3946. [CrossRef] [PubMed]
43. Møller, S.; Yi, X.; Velásquez, S.M.; Gille, S.; Hansen, P.L.M.; Poulsen, C.P.; Olsen, C.E.; Rejzek, M.; Parsons, H.; Yang, Z.; et al. Identification and evolution of a plant cell wall specific glycoprotein glycosyl transferase, ExAD. *Sci. Rep.* **2017**, *7*, 45341. [CrossRef] [PubMed]
44. Miyamoto, T.; Takada, R.; Tobimatsu, Y.; Suzuki, S.; Yamamura, M.; Osakabe, K.; Osakabe, Y.; Sakamoto, M.; Umezawa, T. Double knockout of OsWRKY36 and OsWRKY102 boosts lignification with altering culm morphology of rice. *Plant Sci.* **2020**, *296*, 110466. [CrossRef]
45. Martone, P.T.; Estevez, J.M.; Lu, F.; Ruel, K.; Denny, M.W.; Somerville, C.; Ralph, J. Discovery of lignin in seaweed reveals convergent evolution of cell-wall architecture. *Curr. Biol.* **2009**, *19*, 169–175. [CrossRef] [PubMed]
46. Koutaniemi, S.; Toikka, M.M.; Kärkönen, A.; Mustonen, M.; Lundell, T.; Simola, L.K.; Kilpeläinen, I.A.; Teeri, T.H. Characterization of basic *p*-coumaryl and coniferyl alcohol peroxidases from a lignin-forming *Picea abies* suspension culture. *Plant Mol. Biol.* **2005**, *58*, 141–157. [CrossRef] [PubMed]
47. Quiroga, M.; Guerrero, C.; Botella, M.A.; Barcelo, A.; Amaya, I.; Medina, M.I.; Alonso, F.J.; de Forchetti, S.M.; Tigier, H.; Valpuesta, V. A tomato peroxidase involved in the synthesis of lignin and suberin. *Plant Physiol.* **2000**, *122*, 1119–1127. [CrossRef]
48. Gabaldón, C.; López-Serrano, M.; Pedreño, M.A.; Ros Barceló, A. Cloning and molecular characterization of the basic peroxidase isoenzyme from *Zinnia elegans*, an enzyme involved in lignin biosynthesis. *Plant Physiol.* **2005**, *139*, 1138–1154. [CrossRef] [PubMed]

49. Moural, T.W.; Lewis, K.M.; Barnaba, C.; Zhu, F.; Palmer, N.A.; Sarath, G.; Scully, E.D.; Jones, J.P.; Sattler, S.E.; Kang, C. Characterization of class III peroxidases from switchgrass. *Plant Physiol.* **2017**, *173*, 417–433. [CrossRef]
50. Cochrane, F.C.; Davin, L.B.; Lewis, N.G. The Arabidopsis phenylalanine ammonia lyase gene family: Kinetic characterization of the four PAL isoforms. *Phytochemistry* **2004**, *65*, 1557–1564. [CrossRef]
51. Baltas, M.; Lapeyre, C.; Bedos-Belval, F.; Maturano, M.; Saint-Aguet, P.; Roussel, L.; Duran, H.; Grima-Pettenati, J. Kinetic and inhibition studies of cinnamoyl-CoA reductase 1 from *Arabidopsis thaliana*. *Plant Physiol. Biochem.* **2005**, *43*, 746–753. [CrossRef] [PubMed]
52. Wang, J.P.; Naik, P.P.; Chen, H.C.; Shi, R.; Lin, C.Y.; Liu, J.; Shuford, C.M.; Li, Q.; Sun, Y.H.; Tunlaya-Anukit, S.; et al. Complete proteomic-based enzyme reaction and inhibition kinetics reveal how monolignol biosynthetic enzyme families affect metabolic flux and lignin in *Populus trichocarpa*. *Plant Cell* **2014**, *26*, 894–914. [CrossRef]
53. Solé-Gil, A.; Hernández-García, J.; López-Gresa, M.P.; Blázquez, M.A.; Agustí, J. Conservation of thermospermine synthase activity in vascular and non-vascular plants. *Front. Plant Sci.* **2019**, *10*, 663. [CrossRef]
54. Moghe, G.D.; Last, R.L. Something old, something new: Conserved enzymes and the evolution of novelty in plant specialized metabolism. *Plant Physiol.* **2015**, *169*, 1512–1523. [CrossRef]
55. Xu, Z.; Zhang, D.; Hu, J.; Zhou, X.; Ye, X.; Reichel, K.L.; Stewart, N.R.; Syrenne, R.D.; Yang, X.; Gao, P.; et al. Comparative genome analysis of lignin biosynthesis gene families across the plant kingdom. *BMC Bioinform.* **2009**, *10* (Suppl. 11), S3. [CrossRef]
56. Labeeuw, L.; Martone, P.T.; Boucher, Y.; Case, R.J. Ancient origin of the biosynthesis of lignin precursors. *Biol. Direct.* **2015**, *10*, 23. [CrossRef] [PubMed]
57. Renault, H.; Werck-Reichhart, D.; Weng, J.K. Harnessing lignin evolution for biotechnological applications. *Curr. Opin. Biotechnol.* **2019**, *56*, 105–111. [CrossRef]
58. Benito, B.; Rodriguez Navarro, A. Molecular cloning and characterization of a sodium-pump ATPase of the moss *Physcomitrella patens*. *Plant J.* **2003**, *36*, 382–389. [CrossRef] [PubMed]
59. Kroemer, M.; Dreyer, M.K.; Wendt, K.U. APRV—A program for automated data processing, refinement and visualization. *Acta Crystallogr. D Biol. Crystallogr.* **2004**, *60*, 1679–1682. [CrossRef]
60. Minami, A.; Nagao, M.; Arakawa, K.; Fujikawa, S.; Takezawa, D. Abscisic acid-induced freezing tolerance in the moss *Physcomitrella patens* is accompanied by increased expression of stress-related genes. *J. Plant Physiol.* **2003**, *160*, 475–483. [CrossRef] [PubMed]
61. Richardt, S.; Timmerhaus, G.; Lang, D.; Qudeimat, E.; Corrêa, L.G.G.; Reski, R.; Rensing, S.A.; Frank, W. Microarray analysis of the moss *Physcomitrella patens* reveals evolutionarily conserved transcriptional regulation of salt stress and abscisic acid signaling. *Plant Mol. Biol.* **2010**, *72*, 27–45. [CrossRef] [PubMed]
62. Liu, N.; Zhong, N.Q.; Wang, Q.L.; Li, L.J.; Liu, X.L.; He, Y.K.; Xia, G.X. Cloning and functional characterization of *PpDBF1* gene encoding a DRE-binding transcription factor from *Physcomitrella patens*. *Planta* **2007**, *226*, 827–838. [CrossRef]
63. de Marco, A.; Guzzardi, P.; Jamet, E. Isolation of tobacco isoperoxidases accumulated in cell-suspension culture medium and characterization of activities related to cell wall metabolism. *Plant Physiol.* **1999**, *120*, 371–382. [CrossRef] [PubMed]
64. Chomczynski, P. A reagent for the single-step simultaneous isolation of RNA, DNA and proteins from cell and tissue samples. *BioTechniques* **1993**, *15*, 532–534. [PubMed]
65. Pfaffl, M.W.; Horgan, G.W.; Dempfle, L. Relative expression software tool (REST) for group-wise comparison and statistical analysis of relative expression results in real-time PCR. *Nucleic Acids Res.* **2002**, *30*, e36. [CrossRef] [PubMed]

plants

MDPI

Communication

# Plant Cell Wall Hydration and Plant Physiology: An Exploration of the Consequences of Direct Effects of Water Deficit on the Plant Cell Wall

David Stuart Thompson * and Azharul Islam

School of Life Sciences, University of Westminster, 115 New Cavendish Street, London W1W 6UW, UK; azharul.mywestminster@gmail.com
* Correspondence: s.thompson@westminster.ac.uk

**Abstract:** The extensibility of synthetic polymers is routinely modulated by the addition of lower molecular weight spacing molecules known as plasticizers, and there is some evidence that water may have similar effects on plant cell walls. Furthermore, it appears that changes in wall hydration could affect wall behavior to a degree that seems likely to have physiological consequences at water potentials that many plants would experience under field conditions. Osmotica large enough to be excluded from plant cell walls and bacterial cellulose composites with other cell wall polysaccharides were used to alter their water content and to demonstrate that the relationship between water potential and degree of hydration of these materials is affected by their composition. Additionally, it was found that expansins facilitate rehydration of bacterial cellulose and cellulose composites and cause swelling of plant cell wall fragments in suspension and that these responses are also affected by polysaccharide composition. Given these observations, it seems probable that plant environmental responses include measures to regulate cell wall water content or mitigate the consequences of changes in wall hydration and that it may be possible to exploit such mechanisms to improve crop resilience.

**Keywords:** plant cell wall composition; expansins; water stress; salt stress

**Citation:** Thompson, D.S.; Islam, A. Plant Cell Wall Hydration and Plant Physiology: An Exploration of the Consequences of Direct Effects of Water Deficit on the Plant Cell Wall. *Plants* **2021**, *10*, 1263. https://doi.org/10.3390/plants10071263

Academic Editors: Penélope García-Angulo and Asier Largo-Gosens

Received: 21 May 2021
Accepted: 15 June 2021
Published: 22 June 2021

**Publisher's Note:** MDPI stays neutral with regard to jurisdictional claims in published maps and institutional affiliations.

**Copyright:** © 2021 by the authors. Licensee MDPI, Basel, Switzerland. This article is an open access article distributed under the terms and conditions of the Creative Commons Attribution (CC BY) license (https://creativecommons.org/licenses/by/4.0/).

## 1. Introduction

It is generally accepted that plant cells expand by slow irreversible deformation of their cell walls as a result of stresses generated in the walls by the internal turgor pressure of the cell [1], and thus their growth can be framed as a biomechanical interaction between the properties of the wall of a cell and its turgor pressure. Water stress is expected to affect plant growth as a result of effects on either wall stress or wall mechanical behavior.

The turgor pressure of a cell is determined by its internal osmotic pressure and the water potential outside the cell. For a non-growing cell, it is simply the sum of these terms; however, in a growing cell there must be an imbalance favoring water movement into the cell, although the difference is likely to be slight at a cellular level. Because water potential influences turgor pressure, it is clear that the water status of a plant can alter wall stress and that this can affect the rate of plant growth; however, it should be noted that in many cases the concentration of cellular solutes increases to maintain turgor pressure [2]. There are also many examples of the active modulation of wall mechanical properties in response to signals associated with water stress or water stress itself (e.g., [3]). However, even though water makes up the majority of the primary walls of plant cells by mass (typically >80%, [4]), the direct effects of the water status of a plant on its wall properties have been relatively unexplored. Given the proportion of primary cell walls of plants that water comprises, it would be surprising if the volume that it occupies does not contribute to the mechanical behavior of the wall.

There is good reason to expect that changes in the water content of plant cell walls could affect their mechanical behavior directly. In polymer matrices, the free volume between polymer molecules is thought to have substantial effects on their properties and behavior. This effect is routinely exploited to control the properties of plastics by the addition of the smaller spacing molecules, termed plasticizers (e.g., [5]). Plant cell walls comprise a composite of polymers, with the space between them generally occupied by water (although water may be displaced by other materials in secondary wall deposition).Therefore, water appears analogous to plasticizers in synthetic plastics. For example, pectin films become several-fold stiffer as their degree of hydration is reduced [6].

Tests of the effect of altering the water content on cell walls of rye coleoptiles [7] and sunflower hypocotyls [8] by using osmotica have shown the rate at which walls extended under constant load was reduced as water was removed from the walls and that these effects occurred at water potentials that plants might be expected to experience under "field" conditions (−0.4 MPa in sunflower). The tissue in both sets of experiments had been frozen and thawed, and so the effects of reduced turgor pressure and of cellular responses to the altered water potential could be excluded and the observed effects attributed solely to direct effects of hydration on the cell walls. Substantial reductions in hydration were found in tomato fruit pericarp cell wall material at even more moderate water potentials, with hydration decreasing by more than 50% as the water potential was reduced from 0 MPa to −0.15 MPa [9].

If these effects occur in vivo, changes in wall hydration, spacing and tissue biomechanics as a result of plant water status would be expected to have a direct impact on plant growth. Additionally, reducing the available space in the wall might impede the movement of enzymes and other molecules within the wall [10] and amplify the water potential gradients across tissues if apoplastic hydraulic conductivity decreases because of reduced wall area (e.g., [11]). It would therefore be surprising if responses to water availability do not include measures to maintain wall hydration or to mitigate the consequences of changes. The exploration of such adaptations may shed light on plant responses to water availability and perhaps also offer ways to maintain crop yields and improve resilience when water supplies are limited.

Such measures might include changes in wall polysaccharide composition, structure or charge to alter the relationship between wall water content and water potential, regulation of the osmotic pressure, pH or ion concentrations in the wall space to maintain hydration, or wall "loosening" to minimize any effects of hydration changes on extensibility.

One potential mechanism for controlling the water content of a cell wall is by altering its composition. In principle, this might involve changes in which polysaccharides are present or their relative proportions, but changes in polymer length and the number and composition of branches may also alter these interactions. For example, the quantity of water bound per monosaccharide unit of chitosan, alginate and cellulose in paper differs, as does the way that this relationship changes with water activity [12], such that they dehydrate at different rates as the water potential is reduced. In practice, the pectins in the wall possess the greatest chemical and structural complexity and thus potentially offer multiple mechanisms for modulating these relationships, as does the degree of esterification of pectic acidic groups.

The primary factors determining this relationship between water potential (or water activity) and the quantity of water associated with polymers are the space occupied by the polymer (itself affected by polymer length, both the number and length of branches and the mobility of polymer segments) and the strength of interactions between the polymer functional groups and water [12]. Interactions with groups at chain ends (and therefore the number of chain ends) may also be important [13]. All of these properties could be modulated by alterations in wall biosynthesis or by enzymes acting upon existing wall material.

In addition to the individual interactions between the polymers and water, the cell wall as a whole has properties resulting from interactions between its polymer components, conferring resistance to compression or conversely reducing chain separation or mobility. It certainly seems that a number of wall components have roles in maintaining the spacing required for correct wall function [14]. This "scaffolding" may be necessary to prevent lateral collapse of walls as they deform, but also potentially to prevent compression of walls by the turgor pressures of adjacent cells pushing against one another in tissues exhibiting tissue pressure/tension [15]. Such effects on hydration as a result of interactions between polymers would be expected to include the "egg box" bridges formed by divalent cations between pectic uronic acid groups [16] as well as factors affecting the strength of these interactions, such as ionic strength and pH, if this alters the density of charged groups [9].

Cell wall enzymes offer more rapid mechanisms for modifying the relationship between water potential and wall water content than could be achieved by bulk changes in wall composition. Many types of modification could have such effects, including (but not limited to) changes in polymer molecular weight, degree of polysaccharide branching and length of branches, the number of polymer ends, and the density of charged groups. Another group of cell wall enzymes of considerable interest are the wall loosening proteins known as expansins, because it has been observed that in addition to making plant cell walls and other cellulosic materials more extensible at pH < 5.5 (e.g., [17]), expansins also cause swelling in these materials [18,19]. It therefore seems likely that, in principle, expansins could maintain or increase wall spacing under conditions of water stress.

The following experiments explore the effects of water potential on water content and mechanical behavior and of wall composition and expansins, and the interaction between them, upon these relationships in both plant materials and synthetic models of plant cell walls based on bacterial cellulose. These data establish that the water potentials that plants may experience under field conditions can directly affect the mechanical characteristics of their cell walls in ways that would be expected to affect plant growth and development in vivo and that these effects can be modulated by wall composition and expansin activity.

## 2. Results

In classical experiments, the pore size of plant cell walls was determined by observing whether polyethylene glycols (PEGs) of a range of molecular weights caused plasmolysis or cell collapse, called cytorrhysis [20]. Smaller $M_W$ PEG molecules penetrated the cell wall and exerted an osmotic pressure at the cell membrane, causing plasmolysis, whereas larger $M_W$ PEG molecules were excluded from the wall and caused cytorrhysis. However, osmosis depends on the exclusion of solute, and so these experiments also demonstrated that because PEG molecules with a molecular weight of >4 kDa could not penetrate plant cell walls, they could be used to apply osmotic potentials to them and modify their water content. This method has previously been used to show that plant cell wall material becomes less extensible if PEG is used to reduce its water content in constant-load extensiometer (also known as creep) measurements [7,8,19]. Here, we examine the effects of reduced water potentials generated using PEG on the retention of water and mechanical behavior in bacterial cellulose and composites of bacterial cellulose with pectin and xyloglucan in order to investigate the effects of incorporating these cell wall components.

Figure 1a illustrates examples of the effect of reducing the effective water potential of buffers bathing the bisected frozen and thawed sunflower hypocotyls extending under an applied load. In these experiments, the applied water potential was reduced from 0 MPa to −0.62 MPa or was increased from −0.62 MPa to 0 MPa by exchanging control buffers and buffers containing PEG 6000. Note that the solutes in the buffer should penetrate the wall freely and therefore will not exert an osmotic potential on the wall. Reducing wall water potential, and therefore water content, caused a transient increase in length, followed by a substantial reduction in the long-term rate of extension most analogous to growth in its rate and the period over which it occurs. Increasing the water potential caused a rapid increase in the long-term extension rate after a short lag. It was reasonably common to see a slight shortening of the cell wall material during this lag period (illustrated in detail in Figure 1b, along with the initial increase in length when water content was reduced), and so it seems likely that it mirrors the lengthening during reductions in water content (although of lesser magnitude) and that these effects result from lateral realignment of wall components as the total volume in which they are contained changes.

Sample data from similar experiments using bacterial cellulose and bacterial cellulose composite with apple pectin and tamarind xyloglucan (CPX) are shown in Figure 1c,d respectively. In these experiments the buffers were exchanged twice so that either the control buffer was replaced with buffer containing PEG for a period before returning the strips to the control buffer or the buffer containing PEG was replaced for a period with the control buffer before returning the strips to the buffer containing PEG. As for the sunflower material, the PEG generated an osmotic potential of −0.62 MPa.

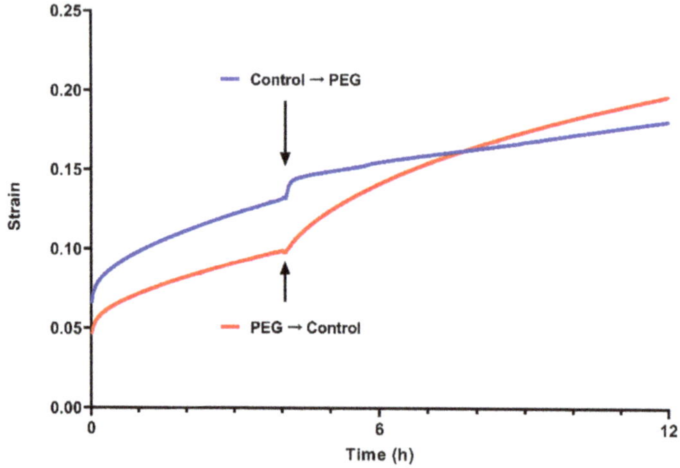


(a)

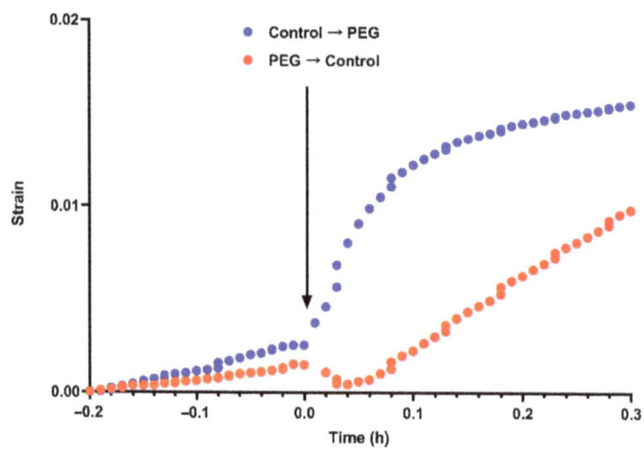


(b)

**Figure 1.** *Cont.*

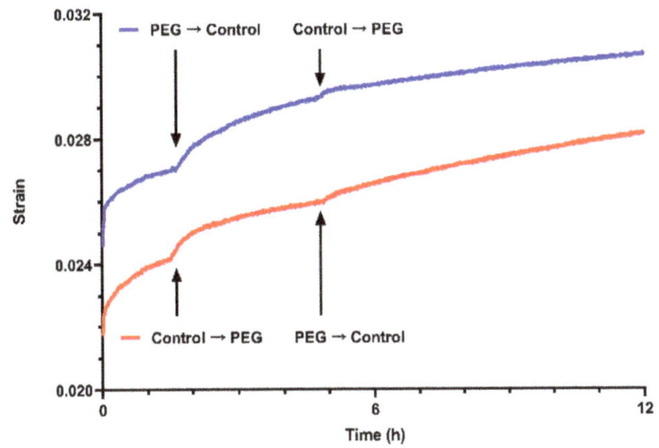


(c)

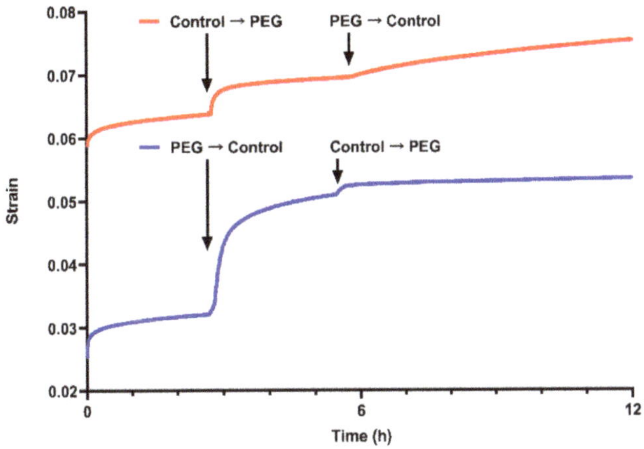


(d)

**Figure 1.** Effects of exchanging MES buffer for the same buffer containing PEG and the opposite exchanges, altering the osmotic potential from 0 MPa to −0.62 MPa or from −0.62 MPa to 0 MPa (**a**) and (**b**) on extension of cell walls from sunflower hypocotyls, (**c**) on bacterial cellulose and (**d**) on bacterial cellulose composite with pectin and xyloglucan. Arrows indicate the times at which osmotic potential was increased or reduced. Strain = $Ln(L_t/L_0)$, where $L_t$ = length at time t and $L_0$ = length at plot time zero (these were lengths immediately before load was increased in plots (**a**), (**c**) and (**d**) and lengths 12 min before buffers were exchanged in (**b**). (**a**) Sunflower hypocotyls extending under an applied load of 0.196 N (applied at 0 h). Blue: 0 MPa → −0.62 MPa; Red: −0.62 MPa → 0 MPa. (**b**) Detail of data in Figure 1a illustrating changes in strain immediately after increasing or decreasing water potential. Blue: 0 MPa → −0.62 MPa; Red: −0.62 MPa → 0 MPa. (**c**) Strips of bacterial cellulose extending under an applied load of 0.196 N (applied at 0 h). Blue: −0.62 MPa → 0 MPa → −0.62 MPa; Red: 0 MPa → −0.62 MPa → 0 MPa. (**d**) Strips of bacterial cellulose composite with pectin and xyloglucan extending under an applied load of 0.098 N (applied at 0 h). Blue: −0.62 MPa → 0 MPa → −0.62 MPa; Red: 0 MPa → −0.62 MPa → 0 MPa.

Although the bacterial cellulose was much less extensible than the sunflower walls, and the changes in rate of extension are therefore less clear than in the plant cell wall material, the same broad effect of water content was observed.

In CPX, long-term extension was substantially reduced when the water content of the material decreased, as was seen in both the plant cell wall material and the bacterial cellulose. However, the material was considerably more extensible than cellulose without pectin and xyloglucan (note that these data are for a lower load than those for the sunflower walls and bacterial cellulose) and exhibited substantial rapid extension after the initial water potential changes, both when the osmotic potential was decreased and when it was increased. The increase in length that occurred after strip hydration was decreased parallels that seen in plant material and in bacterial cellulose alone, although it is considerably enhanced compared to bacterial cellulose. In other experiments, strips of bacterial cellulose composites, which included xyloglucan or other hemicelluloses (including lichenan and mixed-linkage β-glucan), consistently narrowed to a much greater extent than strips of bacterial cellulose or cellulose composite with pectins during constant load extensiometry, suggesting that these hemicelluloses reduced the resistance of the material to lateral compression (in engineering terms, an increase in their Poisson's ratio, which is derived from the ratio of lateral contraction to longitudinal extension as material is stretched (Poisson's ratio for extension in two directions = $-d\varepsilon_{lateral}/d\varepsilon_{longitudinal}$, where $d\varepsilon_{longitudinal}$ is longitudinal strain and $d\varepsilon_{lateral}$ is lateral strain; $d\varepsilon_{lateral}$ is negative because it is a contraction, and so a negative sign is included to give a positive value for the Poisson's ratio).

The extension when the strip water content was increased appears to be a different process (or processes) and can be modelled extremely well ($r^2 > 0.99$) by assuming processes in which an initial extension rate decreases exponentially, reducing by a factor of *e* over a period known as a retardation time (in engineering terms, Kelvin or Kelvin-Voigt elements (a Kelvin element extends with a rate at time t = $r_0(e^{-t/\tau})$, where $r_0$ is the extension rate of the element at time = 0 and $\tau$ = retardation time. Length at time t = $r_0 \times \tau(1 - e^{-t/\tau})$, which is obtained by integrating the previous equation with time). In this case, the model employed two Kelvin elements with retardation times of approximately 5 min and approximately 64 min. These are comparable to those found for creep of plant cell wall material extending under a constant stress [21].

In such rheological models, the individual mathematical elements have often been interpreted as corresponding to different processes by which the material extends (such as the stretching of polymers and realignment of polymers). Therefore, the correspondences with the model of creep in plant material suggests that this "released" extension may indicate equivalent mechanisms of extension. If so, it appears that they were enhanced by incorporation of other wall polysaccharides into the bacterial cellulose matrix. The flow rates of both Kelvin elements in cell walls from tomato fruit epidermis increased at lower pH and were reduced by boiling, indicating that they may have been enhanced by expansins [21]; the pattern of extension after the first increase in water potential also resembles that previously reported in composites of bacterial cellulose treated with expansin [17].

The loss of water from pieces of bacterial cellulose, bacterial composite with apple pectin (CP), and CPX at reduced water potentials was then determined by measuring their change in weight with time after immersion in buffer containing PEG to give a water potential of −0.62 MPa. These data are illustrated in Figure 2a,b, as is their recovery after the pieces were returned to the control buffer without PEG. The proportional changes in weight are summarized in Figure 3. It was consistently observed that none of the materials recovered their original weight and water content completely after they were returned to the buffer that did not contain PEG, exhibiting some degree of hysteresis until they were treated with expansin (the cucumber α-expansin CsExp1 in Figures 2a and 3a) or snail powder extract (in Figures 2b and 3b). Snail acetone powder from the visceral hump of *Helix pomatia* has previously been reported to possess an expansin-like activity and to contain proteins labelled by polyclonal antibodies to purified cucumber expansins [22]; thus, snail powder extracts were used as a second source of expansin activity.

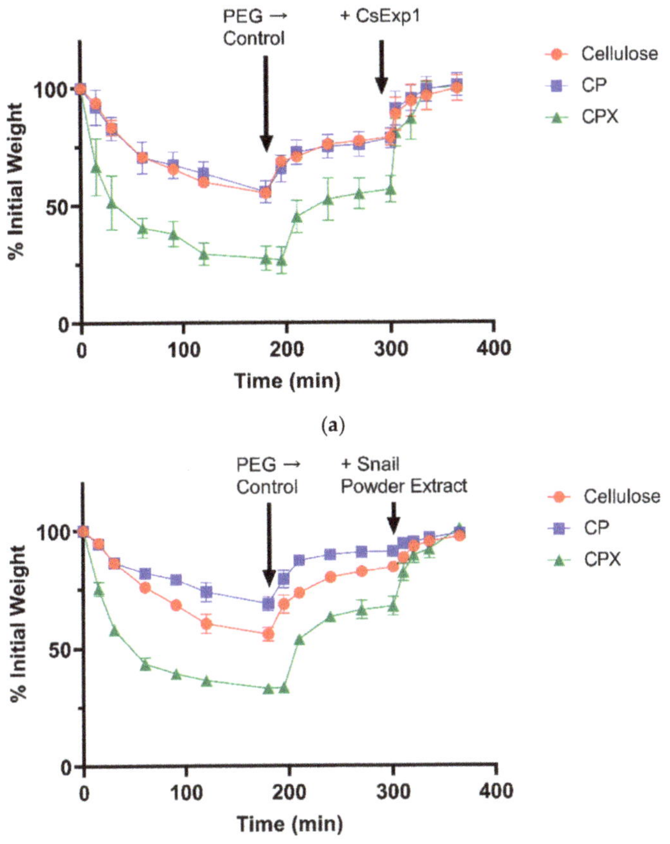


(**a**)

(**b**)

**Figure 2.** Time course of changes in hydration in pieces of bacterial cellulose (red circles), bacterial cellulose composite with apple pectin (blue squares), and bacterial cellulose composite with apple pectin and xyloglucan (green triangles) after the materials were immersed in buffer with an osmotic potential of −0.62 MPa (at 0 h), then returned to control buffer (first arrow), and finally treated with 11.2 µg of α-expansin (second arrow in (**a**)) or 200 µL of snail powder extract (second arrow in (**b**)). Error bars indicate S.D. ($n$ = 3).

Such hysteresis has been seen in other studies of plant cell wall dehydration and rehydration [4]. From these data it seems that space is lost during water removal and reswelling, either because of additional interactions between the cellulose and other components in the composites or because the forces driving reswelling are insufficient to fully restore the original volume, or both, but that these factors can be negated by expansins.

It seems unlikely that that the differences between the materials were because of variations in exclusion of the PEG (at least within the period of these experiments) because the effect of a large but permeating osmoticum (PEG 1000) on sunflower cell walls was initial water loss followed by partial recovery as the osmoticum diffused into the wall (see Appendix A, Figure A1). This was not observed for these cellulose, CP, or CPX batches, where recovery did not occur until after the materials were returned to the control buffer (although it was seen in batches of bacterial cellulose or cellulose composite produced in incubations of less than 120 h).

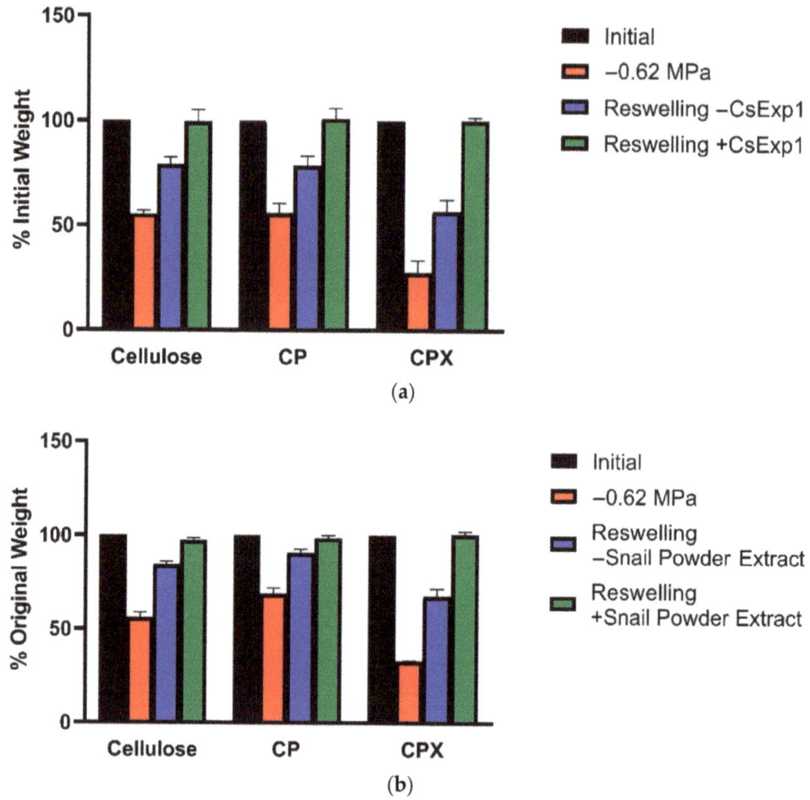


**Figure 3.** Summary of changes in hydration in pieces of bacterial cellulose, bacterial cellulose composite with apple pectin (CP), and bacterial cellulose composite with apple pectin and xyloglucan (CPX) as a percentage of initial weight (black fill), after 180 min at −0.62 MPa (red fill), after 120 min in control buffer (blue fill), and after treatment with 11.2 µg of α-expansin (green fill in (**a**)) or 200 µL of snail powder extract (green fill in (**b**)). Error bars indicate S.D. ($n$ = 3).

Therefore, it seems that incorporation of xyloglucan substantially enhanced the effect of lower water potentials on water content and the magnitude of these changes; this may be reflected in the comparatively large rapid changes in length when water content of CPX strips was increased and decreased for the first time (Figure 1d). There were variations between batches of the cellulose and composites, and the CP batch used for the data in Figures 2b and 3b retained water better than cellulose alone.

In other experiments, the composite produced using polygalacturonic acid (Sigma-Aldrich, cat. No. 81325) lost considerably less water than the composite with apple pectin in PEG solutions, giving an osmotic potential of −0.62 MPa for 180 min (composite with apple pectin lost 34.11 ± 2.26% and composite with polygalacturonic acid lost 20.16 ± 2.61%). This indication that charge density can affect these interactions raises the question of whether apoplastic pH could also affect the wall volume by altering the proportion of pectin uronic acid groups that possess a negative charge (this has been suggested by observation of the effect of pH on hydration of tomato cell wall material [13]). Observations of the effect of pH on the viscosity of apple pectin solutions/gels in the presence of $Ca^{2+}$ from flow-through tests (Figure 4) and of its fluidity in inverted tubes (Appendix A, Figure A2) appear to confirm that the properties of pectins are altered as pH is increased within potentially physiological wall conditions between pH 4 and pH 6, and it seems probable that this is because of $Ca^{2+}$ cross-linking uronic acid groups.

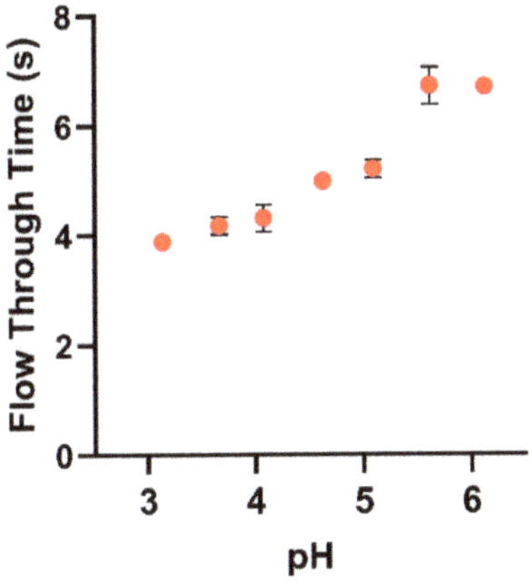


**Figure 4.** The effect of pH on time taken for 2% ($w/v$) solutions of apple pectin of pH 3.14, 3.67, 4.08, 4.63, 5.1, 5.62, and 6.13 containing 5 mM $CaCl_2$ to flow between two graduations of a 1 mL glass pipette. Error bars indicate S.D. ($n$ = 3).

It is possible to observe the effect of expansin on the volume of plant cell walls from an increase in the optical density (O.D.) of suspensions of fragments of cell wall material in buffer. This is illustrated for fragments of wall material prepared from etiolated sunflower hypocotyls in response to 14 µg of CsExp1 and 200 µL of snail powder extract in Figure 5a. The increase is extremely rapid and is largely complete within 2 min of adding expansin. The mean change in optical density 2 min after treatment with CsExp1, snail powder extract horseradish peroxidase (2 units), and hydrogen peroxide (final concentration = 84 mM $H_2O_2$) are shown in Figure 5b. Note that the cell wall material was boiled to inactivate endogenous expansins before homogenization. The effects of peroxidase and hydrogen peroxide, both of which caused slight reductions in O.D., are included for interest, because it has been reported that peroxidative cross-linking of cell wall components such as extensins causes a reduction in cell wall volume [23].

To determine whether cell wall composition could affect this response, suspensions prepared from leaves and petioles of Col-0, *mur1-2*, and *mur2* Arabidopsis plants were treated with CsExp1. Both the *mur-1* and *mur-2* mutants have cell walls with reduced fucose content. MUR-1 encodes the GDP-D-mannose-4,6-dehydratase required for synthesis of GDP-L-fucose [24] and MUR-2 for the fucosyltransferase 1 that fucosylates xyloglucan in Arabidopsis [25]. Therefore, Arabidopsis *mur1* mutants haven substantially reduced fucose content in xyloglucan, rhamnogalacturonan I, rhamnogalacturonan II (RGII), and arabinogalactan proteins, but in *mur2* mutants the reduction is restricted to xyloglucans. Although the leaf cell walls of *mur1* mutants contain normal amounts of RGII, only half forms the borate cross-linked dimer, whereas almost 95% of the RGII forms the borate dimer in wild-type plants [26]. The *mur1* phenotype can be rescued by treatment with elevated levels of boron [26,27].

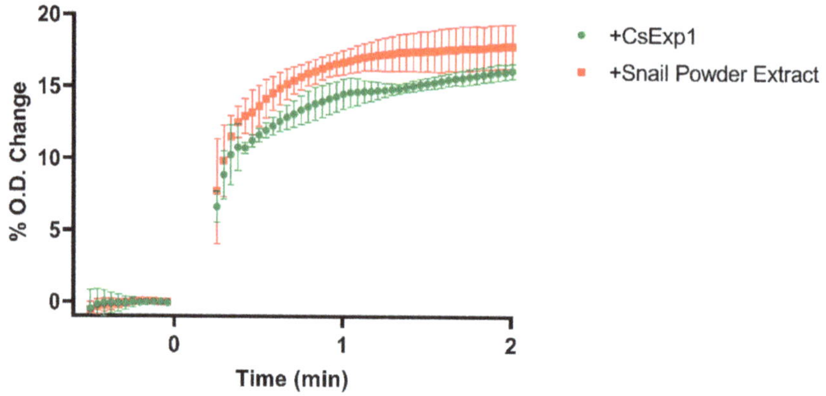


(**a**)

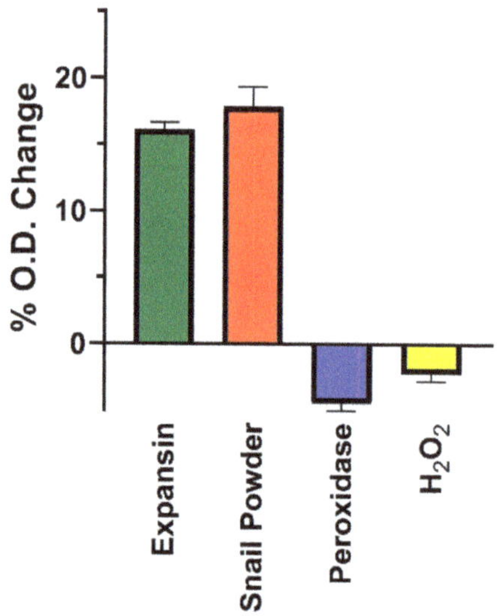


(**b**)

**Figure 5.** Effects of addition of 14 µg CsExp1 expansin or 200 µL snail powder extract on swelling of cell wall fragments from sunflower hypocotyls in suspension, measured by their effect on proportional O.D. at $\lambda$ = 750 nm relative to initial O.D. (**a**) Time course of change in O.D. after addition at 0 s of expansin (green circles) or 200 µL snail powder extract (red squares). (**b**) Summary of O.D. change 2 min after addition of expansin (green fill), snail powder extract (red fill), 2 units of horseradish peroxidase (blue fill), and $H_2O_2$ to give a final concentration of 84 mM $H_2O_2$ (yellow fill). Error bars indicate S.D. ($n$ = 3).

Under the growth conditions used, the Col-0 and *mur2* plants appeared indistinguishable, but the *mur1-2* plants were noticeably smaller. It was also observed that cell wall fragments from *mur1-2* plants exhibited less swelling in response to CsExp1 than those from Col-0 and *mur2* plants (Figure 6a).

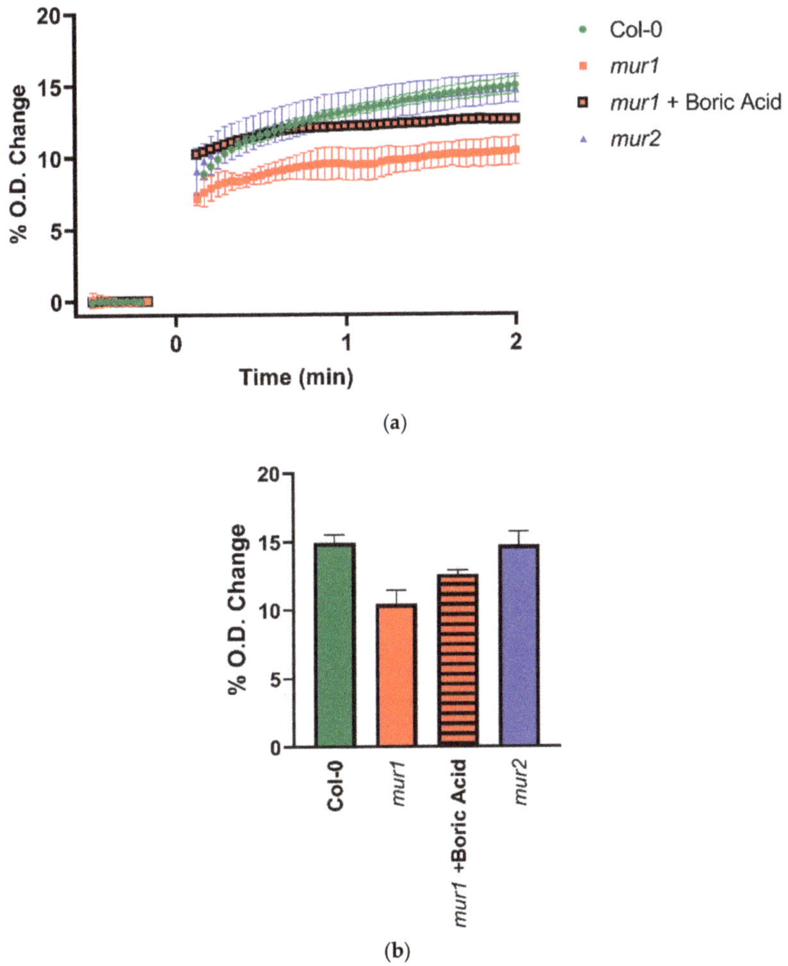


**Figure 6.** Effects of addition of 14 μg CsExp1 expansin on swelling of cell wall fragments from Col-0, *mur1*, and *mur2* Arabidopsis leaves and stems suspended in buffer, measured by its effect on proportional O.D. at $\lambda = 750$ nm relative to initial O.D. (**a**) Time course after addition of expansin at 0 s to Arabidopsis cell wall fragments from wild-type seedlings (green circles), *mur1-2* seedlings (without boric acid pretreatment, red squares; with boric acid pretreatment, red squares with black border) and *mur2* seedlings (blue triangles). (**b**) Summary of the O.D. increase 2 min after addition of expansin to Arabidopsis cell wall fragments from wild-type seedlings (green fill), *mur1-2* seedlings (red fill), fragments from *mur1-2* seedlings pretreated with boric acid (red fill with hatching), and fragments from *mur2* (blue fill). Error bars indicate S.D. ($n = 3$).

No differences were observed between the effect of expansin on cell wall fragments from Col-0 and *mur2*, and pretreating *mur1-2* cell wall fragments with boric acid partially restored swelling in response to expansin (Figure 6b), strongly suggesting that at least part of the reduction in expansin-mediated wall swelling in material from *mur1-2* seedlings was because of its effect on the side chains of RGII and their capacity to form borate cross-linked dimers [26–30]. These data also establish that, in principle, expansin-mediated swelling is influenced by wall composition.

## 3. Discussion

In addition to the well-established impacts of water stress on plant growth as a result of reduced cell turgor pressure and active environmental responses, it is likely that there are also previously unexplored effects of water deficits on cell wall polymer spacing and wall mechanical behavior. Indeed, the function of some observed stress responses may be in part to stabilize wall spacing or to compensate for changes.

As was noted in the introduction, such responses might include changes in wall composition as a result of de novo wall deposition or modification of existing wall components as well as the action of expansins. Additionally, while pH might have a range of effects, it notably affects expansin activity [31]. Ion concentrations could both cause osmotic effects, and in the case of $Ca^{2+}$ ions, mediate cross-linking between pectin uronic acid groups [9].

The results presented demonstrate a profound effect on wall water hydration at reduced water potentials as a result of incorporation of xyloglucan into bacterial cellulose composites; thus, polysaccharide composition can clearly determine relationships between water potential and spacing in materials analogous to plant cell walls. The substantial short-term changes in extension and changes in hydration in CPX compared to bacterial cellulose alone hint that (at least in these materials) xyloglucan, and perhaps other hemicelluloses, facilitate realignment of the cellulose under applied forces, whether extensive (from the extensiometric data) or compressive (from the effects of water potential on water content).

However, as a result, incorporating xyloglucan into bacterial cellulose composites substantially increased the effect of water potential on their degree of hydration, and so (unless their effects are markedly different in plant cell walls) it seems likely that they (and perhaps other hemicelluloses) are more likely to accentuate the effects of water potential on spacing than to mitigate them (although if they facilitate realignment of wall components, this may mitigate the mechanical effects of reduced spacing).

Perhaps the most obvious candidates for the role of maintaining hydration are the pectins. Pectins are amongst the most structurally complex of the cell wall polysaccharides and additionally carry varying charge density and distribution, depending on the degree and distribution of esterification of the uronic acid groups. Therefore, pectic interactions with one another and other wall components, variation in their size, degree, and nature of branching, and the effects of ion concentrations and pH on them offer multiple mechanisms for modulating hydration of the wall. It has indeed been found that specific modification of rhamogalacturonan I (RGI) structure by introduction of a β-1,4-galactanase into potato tubers increased the rate of water loss compared to controls and made the material more brittle in mechanical compression tests [32]. We found that cellulose composites with polygalacturonic acid that has been de-esterified retained a greater degree of hydration at an osmotic potential of −0.62 MPa than composites with apple pectin, suggesting that cross-linking can also strengthen the material against collapse. These experiments were in the presence of $Ca^{2+}$ ions, but incorporation of pectins into bacterial cellulose composites increased their resistance to compression even without $Ca^{2+}$, though they became less stiff to shear oscillation [33].

Charged uronic acid groups could affect the relationship between volume and water potential in a number of ways, including charge repulsion, cross-linking via divalent metal ions [34,35], and generating osmotic pressures from cations immobilized by charge interactions (a phenomenon known as the Donnan effect [35]). Hydration of tomato fruit cell wall material was found to decrease with increasing concentration of NaCl (with the greatest change between 0 and 2.5 mM) in 6 mM $CaCl_2$ and in buffers with pH > 4.5 [9]. These observations suggest a complex relationship between pectic charge and volume affected by Donnan effects at low ionic concentration as well as cross-linking by divalent ions (most importantly $Ca^{2+}$ in vivo). Increasing pH led to a reduction in wall swelling, within ranges that might be expected in vivo (between pH 4 and pH 5). This is higher than might have been expected from the $pK_a$ of polygalacturonic (which is typically reported as <3.5; [9]) but is in accord with the data presented in Figures 4 and A2. $Ca^{2+}$ cross-links form cooperatively [36], and so effective cross-linking may depend on the probability that there are clusters of adjacent de-esterified uronic groups, "shifting" the pH at which cross-linking is affected into physiological ranges.

### 3.1. Water Stress and Cell Wall Composition

If there is a direct effect of plant water potential on wall properties, it seems to be worth examining examples of reported changes in cell wall composition, gene expression, and enzyme quantity or activity in response to water stress for possible evidence of regulation of wall hydration and spacing.

Changing wall composition in response to the reduction of water availability has been reported in a number of instances. Several-fold increases in labelling of xyloglucan and unesterified pectin have been reported during dehydration and rehydration in the resurrection plant *Craterostigma wilmsii* [37], and an increase in arabinose-rich pectins and arabinogalactan proteins have been found in a range of other resurrection plant species [38]. Less extreme examples include an increase in pectic branching in response to a water stress of −0.5 MPa in walls of root tips of a drought tolerant cultivar of wheat, which was not seen in a susceptible cultivar [39], and changes in the polysaccharide composition and molecular weight distribution in cell walls of wheat coleoptiles in response to PEG-induced water stress [40] (although the osmotic potential is not given, we estimate that it would be approximately −0.5 MPa for the concentration of PEG specified). Of these, increasing pectin branching is certainly one type of response that would be expected to maintain wall hydration and spacing.

Researchers have also reported a range of changes in the activity of enzymes capable of acting upon plant cell walls in response to water or osmotic stresses, and in many cases these enzymes would be expected to be involved in remodeling the walls. For example, it seems that xyloglucan transglycosylase-hydrolases (XTHs) and xyloglucan endotransglycosylases (XETs), enzymes involved in modifying xyloglucan chain length, often increase under these conditions. XET activities were found to increase in maize roots close to the tip, where growth was maintained at extreme water deficit conditions (−1.6 MPa), and both growth maintenance and increased XET activity required ABA [41]. Similar increases in XET have been observed in elongating regions of roots of wheat seedlings [42] and in leaves of *Lolium temulentum* L. seedlings [3] experiencing water deficit. Evidence that XTHs are involved in the adaptation of plant cell walls to conditions of water deficit comes from the observation that expression of an XTH gene induced by a range of abiotic stresses in hot pepper conferred resistance to drought and salt stresses in tomato [43].

Whether such changes in XTH or XET activity will indeed affect wall water retention under water stress is, at present, speculative and would be expected to depend on their exact effects on chain length and number of ends. However, the data in Figures 1 and 2 suggest that xyloglucan can have a profound effect on these relationships.

### 3.2. Expansins and Wall Spacing

It is clear that expansins can increase wall spacing, and so the many reported cases of expansins increasing under drought stress are striking. For example, changes in expression in a range of expansins were found in potato in response to ABA and drought (as well as a number of other plant hormones and abiotic stresses), with some genes being upregulated and some downregulated [44]. Expansin activity and transcript levels of a number of expansin genes were also observed to increase in response to a reduced water potential (−1.6 MPa) in the regions of maize root tips where growth continued at these considerable water stresses [45,46]. ABA and osmotic stress generated using PEG increased expression of one expansin (EXP1) in elongating tissues of soybean during germination [47] and both expansin activity and expression of one expansin gene (TaEXPA1) under moderate osmotic stress in wheat leaves, an effect that was markedly greater in a drought-resistant cultivar than one that is more susceptible to drought [48].

These changes could be to increase wall extensibility and thereby compensate for a reduction in cell wall stress due to decreased turgor, but in many cases plant cells increase cell osmotic pressure under conditions of reduced water availability so that turgor pressure and wall stress are substantially unaffected [2,49,50] (although this is not always the case, e.g., [51]). Another hypothetical possibility might be that expansin quantities increase to offset a reduction in activity caused by the apoplastic alkalinization often associated with water stress [52]. However, the data in Figures 2, 3, 5 and 6 show that expansins can increase and restore cell wall spacing and that the magnitude of this effect depends on wall composition. If this is the case, and reduced water potentials affect cell wall spacing, it appears at least plausible that the functions of expansins under conditions of water stress include maintenance or restoration of wall space or mitigation of the effects of spatial constraint on wall stiffening, or both.

Spacing between cell wall components may also be lost as they realign during wall deformation, and given the observed effect of expansins on hysteresis during dehydration and rehydration cycles, it is conceivable that they play a role in controlling the strain hardening that this could cause.

### *3.3. RGII*

RGII comprises a very small proportion of plant cell walls [53], but the *mur1* Arabidopsis mutants, in which its structure is altered, have a drastically dwarfed phenotype. The *mur1* phenotype can be substantially rescued by supplying boric acid; it seems that this is because of a reduction in the borate diester cross-links formed between RGII domains via apiose residues [27].

In the experiments reported here, we found that the swelling of cell wall fragments from *mur1-2* in response to expansin was inhibited compared to wall fragments from wild-type plants. Partial restoration of the response by pre-treatment with boric acid strongly suggests that the difference between wild-type and *mur1* material was because of altered RGII cross-linking. However, cell wall swelling and increased porosity have been reported to result from boron deficiency [54,55].

This apparent contradiction could be because these experiments tested the change in swelling of wall fragments rather than their absolute size, and so the reduced effect of expansin could be because they were already more swollen than those from the control seedlings. However, the observation that salt stress causes root cells to burst in *mur1* [56] and some other Arabidopsis mutants with altered RGII structure [57] suggests the possibility that under some conditions wall swelling might lead to excessive plasticization of the walls. $Na^+$ ions might cause such an effect if they become immobilized by interaction with uronic acid groups and generate an osmotic pressure in the wall (a Donnan effect) or by competing with $Ca^{2+}$ ions for uronic acid groups, thereby reducing cross-linking between pectins other than RGII [58].

Partial restoration of the expansin response by pre-treatment with boric acid strongly suggests that the difference between the effect of expansin on wild-type (and *mur2*) and *mur1* material was because of altered RGII cross-linking. This may have been because the fragment size was reduced by the pre-treatment, giving a lower baseline so that expansion to the same final fragment size gave a greater proportional effect (it is hard to be sure of this because some O.D. was always lost during the washing steps used to remove the boric acid), but it may be worth considering whether RGII cross-linking also contributes to the force required to drive cell expansin-mediated wall swelling and whether RGII cross-linking may mediate a balance between compressive and expansive forces in plant cell walls.

### *3.4. Summary*

It appears that plant cell walls are plasticized by water and that water potentials experienced under field conditions can alter wall water content to a degree that could affect plant growth and development via this mechanism. If this is the case, then it is to be expected that adaptations to conditions of low water availability or environmental fluctuations in water availability will include measures to maintain wall hydration or mitigate the effects of water potential on wall spacing.

We believe that this model offers a novel perspective for interpreting the role of the cell wall in plant adaptations to water stresses. A better understanding of such relationships between plant cell wall composition, hydration, and biomechanics may shed light on a relatively unexplored aspect of plant responses to their environment and offer new targets for improving crop resilience in the Twenty-first century.

## 4. Materials and Methods

### 4.1. Plant Material

Sunflower seeds (Giant Yellow variety, Suttons seeds, Paignton, UK) were imbibed in tap water overnight and then sown in pots containing water-saturated perlite (Silvaperl, Gainsborough, UK). The plants were grown for 5 days at about 30 °C in pots covered with a plastic pot lid and wrapped with aluminum foil to obtain etiolated hypocotyls.

*Arabidopsis thaliana* wild type (Col-0), *mur-1-2* (N6244), and *mur-2* (N8565) seeds (all from The Nottingham Arabidopsis Stock Centre, University of Nottingham, Nottingham, UK) were surface sterilized in 70% ethanol for 2 min and 5% sodium hypochlorite for 10 min, rinsed five times in distilled water, and then spread onto plates of half strength Murashige and Skoog medium supplemented with 1% sucrose and 0.7% agar. The plates were kept at 4 °C for 4 days in complete darkness for stratification and then grown for about 14 days under a 16 h light/8 h dark photoperiod and 23 °C/19 °C temperature regime.

### 4.2. Bacterial Cellulose and Bacterial Cellulose Composites

Bacterial cellulose was produced by culturing *Komagataeibacter sucrofermentans* (ATCC® 700178™, Promochem, London, UK) in 100 mL YGC medium (50 g $L^{-1}$ glucose, 5.0 g $L^{-1}$ yeast extract, 12.5 g $L^{-1}$, $CaCO_3$) in 250 mL Erlenmeyer flasks in static conditions for at least 120 h at 26 °C. A long incubation was necessary to ensure that the pellicles were dense enough to exclude PEG of Mw 6000 (PEG 6000). Composites of bacterial cellulose with pectin were produced by the addition of 0.5% ($w/v$) pectin to the YGC medium, and composites of bacterial cellulose with pectin and xyloglucan by addition of 0.25% ($w/v$) pectin and 0.25% ($w/v$) xyloglucan to the YGC medium (apple pectin, Sigma-Aldrich, Dorset, UK, cat. No. 76282, 70–75% esterified; tamarind seed xyloglucan, Megazyme International, Ireland, Cat. No. P-XYGLN).

### 4.3. Expansin and Snail Powder Extract

Expansin was a gift from Professor Simon McQueen-Mason (University of York, York, UK); it was the $\alpha$-expansin CsExp1 expressed in tomato plants and dissolved in pH 4.5 35 mM MES/35 mM Acetate buffer containing 200 mM NaCl [59].

Snail powder extract was prepared by mixing 50 mg of snail acetone power from the visceral hump of *Helix pomatia* (Sigma-Aldrich, Poole, Dorset, UK, Cat. No. S9764, now discontinued) into 1 mL of MES buffer 10 mM MES buffer containing 5 mM KCl and 1 mM $CaCl_2$ titrated to pH 5.0 with 1 M NaOH using a laboratory vortexer and incubating for 10 min at room temperature before centrifuging for 10 min at 7200× $g$ and collecting the supernatant.

### 4.4. Extensiometry

Segments of 20 mm were cut from the top growing part of the sunflower hypocotyls, longitudinally bisected using a double-edged razor blade, and then frozen using freezing spray (R.A. Lamb, Eastbourne, UK). After 60 s, the segments were thawed in MES buffer. The segments were then pressed between microscope slides wrapped with absorbent paper, using about 2 kg of weight for 60 s, and returned to MES buffer. Segments of cellulose or cellulose composite 2 mm × 1 mm in cross-section and about 10 mm in length were cut from pellicles using a pair of razor blades held in a spaced block and transferred to MES buffer.

Constant load extensiometry was carried out as described in [21]. Segments of plant material, bacterial cellulose, or cellulose composite were fixed into a reservoir tube with about 10 mm exposed between a clamp at the bottom of the tune and another clamp attached to one arm of a counterweighted cantilever. The reservoir tube was then filled with buffer. The exact initial length of the exposed portion of the segment was measured using a magnifying eyepiece with a graticule (extensiometer custom built by the Biological Sciences workshop at the Lancaster University, Lancaster, UK).

The counterweight positions were adjusted so that a negligible force was applied to the samples for 20 min to allow them to settle, after which the force was increased in increments of 0.098 N (10 g) by sliding the counterweights along the cantilever arms. The change in length was measured using a linear variable displacement transducer (Schlumberger DFG 2.5 from RS Components Ltd., Corby, Northants, UK), which gave a voltage output based on the position of a core balanced on the opposite end of the cantilever arm in relation to the one to which the clamped segment was attached. The mean of 1000 individual readings was recorded every 30 s by a Bede PCADH24 analogue input card (Bede Technology, Jarrow, Tyne and Wear, UK). The extensiometer can be used to test up to six samples at a time. All extensiometry experiments were conducted at least twice, using at least two replicates of each treatment in each run. The data presented are examples of single tests, because although the replicate data exhibited the same patterns of response, they are not normally closely enough matched for pooling of multiple runs to be feasible.

The buffer used throughout (referred to as MES buffer) was 10 mM MES buffer containing 5 mM KCl and 1 mM $CaCl_2$, titrated to pH 5.0 with 1 M NaOH. The pH of buffers containing PEG or other osmotica was adjusted to pH 5.0 after their addition. The effect of changing the water potential on segment extension was observed by draining MES buffer from the extensiometer reservoirs and replacing it with the same buffer containing PEG 6000 or by carrying out the opposite exchange. PEG osmotic pressure (and therefore water potential) was determined by vapor pressure osmometry using a Vapor® 5520 osmometer (Wescor Inc., South Logan, UT, USA). Melting point depression measurements of the osmotic pressures of PEG solutions are higher than those made using vapor pressure [60], but it has been plausibly argued that this is because of the effect of temperature on higher order virial coefficients of the Van't Hoff equation in PEG solutions and therefore that measurements from melting point, which are necessarily at low temperatures, do not accurately reflect the osmotic pressure under the measurement conditions [61], and vapor pressure measurements were preferred. From these data, 0.27 g PEG 6000 g water$^{-1}$ gave an osmotic potential of $-0.62$ MPa.

### 4.5. Viscosity of Pectin Solutions

The viscosity of 2% ($w/v$) solutions of apple pectin (Sigma-Aldrich, cat. No. 76282, 70–75% esterified) containing 5 mM $CaCl_2$ adjusted to the required pH by dropwise addition of 2 M KOH were tested by timing how long the solutions took to flow between two graduations of a 1 mL graduated glass pipette. The same pipette was used throughout for consistency and was wetted with the pectin solutions before measurements were carried out. The viscosity of 2% ($w/v$) solutions of apple pectin (Sigma-Aldrich, cat. No. 76282, 70–75% esterified), containing 10 mM $CaCl_2$ adjusted to the required pH by dropwise addition of 2 M KOH, were also visualized by inverting capped disposable centrifuge tubes for 5 min. Measurements were carried out at room temperature.

### 4.6. Measurement of the Effects of Water Potential on Water Content of Bacterial Cellulose and Cellulose Composites

Segments of bacterial cellulose or cellulose composite, 2 mm × 1 mm in cross-section and approximately 10 mm in length, were cut from the cellulose and cellulose composites with a pair of razor blades held in a spaced block and transferred to MES buffer for at least 10 min at room temperature. Segments were then blotted quickly using absorbent paper to remove surface buffer, and their initial weight measured quickly using a microbalance. The pieces were then transferred into Eppendorf tubes containing 800 μL of MES buffer containing PEG 6000 solution (0.27 g PEG 6000 g water$^{-1}$, giving an osmotic potential of −0.62 MPa, determined using a Vapor® 5520 osmometer as noted above) and the weight changes were measured at regular intervals for 180 min by removing the segments from the buffer, quickly rinsing them in MES buffer without PEG to remove the viscous PEG solutions, blotting them to remove surface liquid, and again weighing them before returning them to the tubes of MES buffer with PEG. After 180 min, the pieces were transferred to Eppendorf tubes containing 800 μL of control MES buffer without PEG to rehydrate, and the weight changes again were recorded at regular intervals. Finally, 11.2 μg of α-expansin or 200 μL of snail powder extract was added to the MES buffer, and any further changes in weight were recorded.

### 4.7. Cell Wall Swelling Measurements from Optical Density

Cell wall fragments suspended in buffer were prepared from approximately 1 g of segments from the top 2 cm of sunflower hypocotyls or approximately 1 g of Arabidopsis seedling leaves and stems. The material was boiled for 90 s in distilled water to inactivate endogenous expansins in the plant tissues and then homogenized in 10 mL of MES buffer using a laboratory mixer/emulsifier at full speed for 5 min (Silverson Machines Ltd., Waterside, UK). The homogenized suspension was then centrifuged for 2.5 min at 650× *g* to remove large pieces of material and then centrifuged again for 10 min at 1450× *g* to pellet smaller cell wall fragments of cell wall. These were re-suspended in 10 mL MES buffer. Fragments from Arabidopsis seedlings were pelleted by centrifugation at 1450× *g* and resuspended in 10 mL MES buffer two further times to eliminate pigmentation in the suspensions (although there was negligible absorbance from them at a wavelength of 750 nm).

The optical density of the fragments suspended in buffer at a wavelength of 750 nm was determined using a spectrophotometer (Lambda 20, Perkin Elmer, Waltham, MA, USA). The initial optical density of 1.0 mL in a disposable plastic cuvette was recorded for 30 s, after which expansin (CsExp1) or snail powder extract was added to the suspension. The cuvette was then gently inverted to mix the solution and returned to the spectrophotometer, and the optical density was recorded for a further 120 s. The effects of horseradish peroxidase (Sigma-Aldrich, P6140) and $H_2O_2$ on the optical density of suspensions from sunflower hypocotyls were also tested.

### 4.8. Statistical and Mathematical Analysis

Where error bars are given, experimental data are presented as mean ± standard deviation, calculated using Microsoft Office Excel (Microsoft Corporation, Redmond, WA, USA). Strain at time t relative to time 0 are true strains calculated from $Ln(L_t/L_0)$, where $L_t$ is the segment length at time t and $L_0$ is the length at time 0 (time 0 was immediately before the stress applied to the segment was increased unless otherwise stated). Modeling and parameterization of extensiometric data used the Microsoft Office Excel solver to minimize the squares of differences between modeled and experimental data.

**Author Contributions:** All experimental work except that presented in Figures 5 and A1 was carried out by A.I. under the supervision of D.S.T. The data presented in Figures 5 and A1 was obtained by D.S.T. Conceptualization: D.S.T.; methodology, D.S.T. and A.I.; software, D.S.T.; formal analysis, D.S.T. and A.I.; investigation, A.I.; resources, D.S.T.; data curation, D.S.T. and A.I.; writing—original draft preparation, D.S.T.; writing—review and editing, D.S.T.; visualization, D.S.T. and A.I.; supervision, D.S.T.; project administration, D.S.T.; funding acquisition, D.S.T. All authors have read and agreed to the published version of the manuscript.

**Funding:** This research received no external funding. A.I. received a Cavendish Research PhD Scholarship from the University of Westminster.

**Institutional Review Board Statement:** Not applicable.

**Informed Consent Statement:** Not applicable.

**Data Availability Statement:** Data is contained within the article.

**Acknowledgments:** We thank Simon McQueen-Mason of the University of York for his donation of the CsExp1 used in these experiments. The extensiometer was constructed by Philip Smith at Lancaster University.

**Conflicts of Interest:** The authors declare no conflict of interest. The funders had no role in the design of the study; in the collection, analyses, or interpretation of data; in the writing of the manuscript, or in the decision to publish the results.

## Appendix A

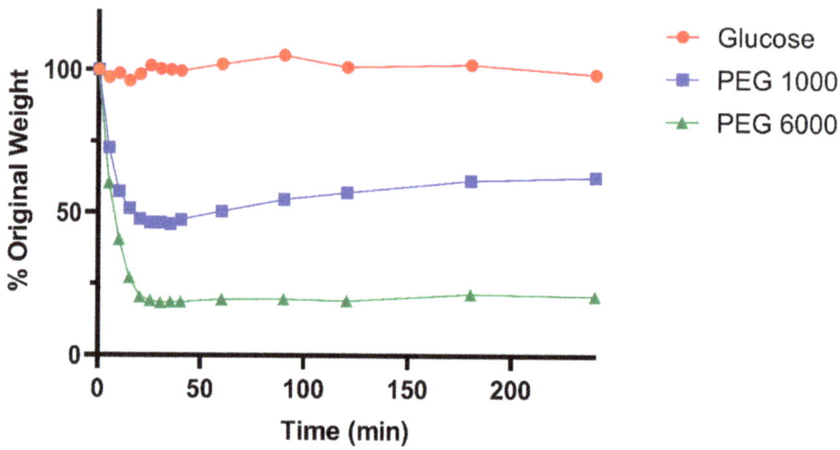


**Figure A1.** Time course of weight changes in cell walls from sunflower hypocotyls exposed to a water potential of −0.62 MPa generated by an osmoticum able to permeate the wall rapidly (glucose, red circles), an osmoticum partly excluded from the wall but expected to diffuse through it more slowly than water (PEG 1000, blue squares), and an osmoticum substantially excluded from the wall (PEG 6000, green triangles).

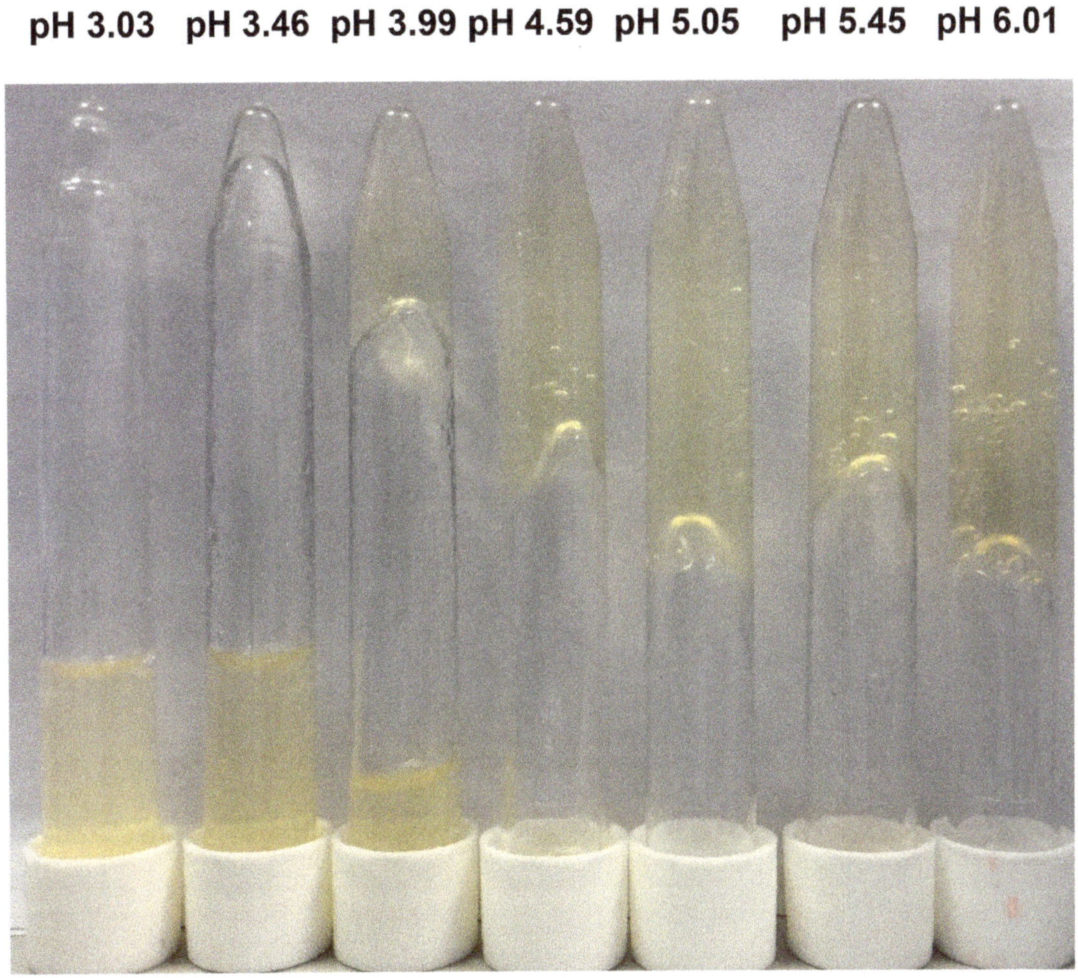


**Figure A2.** Image of tubes containing 2% (*w*/*v*) solutions of apple pectin of pH 3.03, 3.46, 3.99, 4.59, 5.05, 5.45, and 6.01 containing 10 mM $CaCl_2$ five minutes after the tubes were inverted.

## References


1. Braidwood, L.; Breuer, C.; Sugimoto, K. My body is a cage: Mechanisms and modulation of plant cell growth. *New Phytol.* **2014**, *201*, 388–402. [CrossRef]
2. Turner, N.C. Turgor maintenance by osmotic adjustment: 40 years of progress. *J. Exp. Bot.* **2018**, *69*, 3223–3233. [CrossRef]
3. Thompson, D.S.; Wilkinson, S.; Bacon, M.A.; Davies, W.J. Multiple signals and mechanisms that regulate leaf growth and stomatal behaviour during water deficit. *Physiol. Plant.* **1997**, *100*, 303–313. [CrossRef]
4. Hardegree, S.P. Derivation of plant cell wall water content by examination of the water-holding capacity of membrane-disrupted tissues. *J. Exp. Bot.* **1989**, *219*, 1099–1104. [CrossRef]
5. Fong, R.J.; Robertson, A.; Mallon, P.E.; Thompson, R.L. The impact of plasticizer and degree of hydrolysis on free volume of poly(vinyl alcohol) films. *Polymers* **2018**, *10*, 1036. [CrossRef] [PubMed]
6. Zsivanovits, G.; MacDougall, A.J.; Smith, A.C.; Ring, S.G. Material properties of concentrated pectin networks. *Carbohyd. Res.* **2004**, *339*, 1317–1322. [CrossRef] [PubMed]
7. Edelmann, H.G. Water potential modulates extensibility of rye coleoptile cell walls. *Bot. Acta* **1995**, *108*, 374–380. [CrossRef]
8. Evered, C.; Majevadia, B.; Thompson, D.S. Cell wall water content has a direct effect on extensibility in growing hypocotyls of sunflower (*Helianthus annuus* L.). *J. Exp. Bot.* **2007**, *58*, 3361–3371. [CrossRef]

9. MacDougall, A.J.; Rigby, N.M.; Ryden, P.; Tibbits, C.W.; Ring, S.G. Swelling behavior of the tomato cell wall network. *Biomacromolecules* **2001**, *2*, 450–455. [CrossRef] [PubMed]
10. Passioura, J.B. The physical chemistry of the primary cell wall: Implications for the control of expansion rate. *J. Exp. Bot.* **1994**, *45*, 1675–1682. [CrossRef]
11. Boanares, D.; Ferreira, B.G.; Kozovits, A.R.; Sousa, H.C.; Isaias, R.M.S.; França, M.G.C. Pectin and cellulose cell wall composition enables different strategies to leaf water uptake in plants from tropical fog mountain. *Plant Physiol. Bioch.* **2018**, *122*, 57–64. [CrossRef] [PubMed]
12. Despond, S.; Espuche, E.; Cartier, N.; Domard, A. Hydration mechanism of polysaccharides: A comparative study. *J. Polym. Sci. B Polym. Phys.* **2005**, *43*, 48–58. [CrossRef]
13. Gocho, H.; Shimizu, H.; Tanioka, A.; Chou, T.-J.; Nakajima, T. Effect of polymer chain end on sorption isotherm of water by chitosan. *Carbohyd. Polym.* **2000**, *41*, 87–90. [CrossRef]
14. Jones, L.; Milne, J.L.; Ashford, D.; McQueen-Mason, S.J. Cell wall arabinan is essential for guard cell function. *Proc. Natl. Acad. Sci. USA* **2003**, *100*, 11783–11788. [CrossRef] [PubMed]
15. McClendon, J.H. The balance of forces generated by the water potential in the cell-wall-matrix: A model. *Am. J. Bot.* **1981**, *68*, 1263–1268. [CrossRef]
16. Braybrook, S.A.; Hofte, H.; Peaucelle, A. Probing the mechanical contributions of the pectin matrix. *Plant Signal. Behav.* **2012**, *7*, 1037–1041. [CrossRef]
17. Whitney, S.E.C.; Gidley, M.J.; McQueen-Mason, S.J. Probing expansin action using cellulose/hemicellulose composites. *Plant J.* **2000**, *22*, 327–334. [CrossRef] [PubMed]
18. Yennawar, N.H.; Li, L.-C.; Dudzinski, D.M.; Tabuchi, A.; Cosgrove, D.J. Crystal structure and activities of EXPB1 (Zea m 1), a β-expansin and group-1 pollen allergen from maize. *Proc. Natl. Acad. Sci. USA* **2006**, *103*, 14664–14671. [CrossRef] [PubMed]
19. Thompson, D.S. Space and Time in the Plant Cell Wall: Relationships between Cell Types, Cell Wall Rheology and Cell Function. *Ann. Bot.* **2008**, *101*, 203–211. [CrossRef]
20. Carpita, N.; Sabularse, D.; Montezinos, D.; Delmer, D.P. Determination of the pore size of cell walls of living plant cells. *Science* **1979**, *205*, 1144–1147. [CrossRef]
21. Thompson, D.S. Extensiometric determination of the rheological properties of the epidermis of growing tomato fruit. *J. Exp. Bot.* **2001**, *52*, 1291–1301. [CrossRef] [PubMed]
22. Cosgrove, D.J.; Durachko, D.M. Autolysis and extension of isolated walls from growing cucumber hypocotyls. *J. Exp. Bot.* **1994**, *45*, 1711–1719. [CrossRef]
23. Pereira, C.S.; Ribeiro, J.M.L.; Vatulescu, A.D.; Findlay, K.; MacDougall, A.J.; Jackson, P.A.P. Extensin network formation in *Vitis vinifera* callus cells is an essential and causal event in rapid and $H_2O_2$-induced reduction in primary cell wall hydration. *BMC Plant Biol.* **2011**, *11*, 106. [CrossRef]
24. Bonin, C.P.; Potter, I.; Vanzin, G.F.; Reiter, W.-D. The MUR1 gene of *Arabidopsis thaliana* encodes an isoform of GDP-d-mannose-4,6-dehydratase, catalyzing the first step in the *de novo* synthesis of GDP-l-fucose. *Proc. Natl. Acad. Sci. USA* **1997**, *94*, 2085–2090. [CrossRef] [PubMed]
25. Vanzin, G.F.; Madson, M.; Carpita, N.C.; Raikhel, N.V.; Keegstra, K.; Reiter, W.-D. The *mur2* mutant of *Arabidopsis thaliana* lacks fucosylated xyloglucan because of a lesion in fucosyltransferase AtFUT1. *Proc. Natl. Acad. Sci. USA* **2002**, *99*, 3340–3345. [CrossRef]
26. O'Neill, M.A.; Eberhard, S.; Albersheim, P.; Darvill, A.G. Requirement of borate cross-linking of cell wall rhamnogalacturonan II for Arabidopsis growth. *Science* **2001**, *294*, 846–849. [CrossRef] [PubMed]
27. O'Neill, M.A.; Ishii, T.; Albersheim, P.; Darvill, A.G. Rhamnogalacturonan II: Structure and function of a borate cross-linked cell wall pectic polysaccharide. *Ann. Rev. Plant Biol.* **2004**, *55*, 109–139. [CrossRef]
28. Reiter, W.D.; Chapple, C.; Somerville, C.R. Mutants of Arabidopsis thaliana with altered cell wall polysaccharide composition. *Plant J.* **1997**, *12*, 335–345. [CrossRef]
29. Ryden, P.; Sugimoto-Shirasu, K.; Smith, A.C.; Findlay, K.; Reiter, W.-D.; McCann, M.C. Tensile properties of Arabidopsis cell walls depend on both a xyloglucan cross-linked microfibrillar network and rhamnogalacturonan II-borate complexes. *Plant Physiol.* **2003**, *132*, 1033–1040. [CrossRef] [PubMed]
30. Reuhs, B.L.; Glenn, J.; Stephens, S.B.; Kim, J.S.; Christie, D.B.; Glushka, J.G.; Zablackis, E.; Albersheim, P.; Darvill, A.G.; O'Neill, M.A. L-Galactose replaces L-fucose in the pectic polysaccharide rhamnogalacturonan II synthesized by the L-fucose-deficient *mur1 Arabidopsis* mutant. *Planta* **2004**, *219*, 147–157. [CrossRef]
31. McQueen-Mason, S.; Durachko, D.M.; Cosgrove, D.J. Two endogenous proteins that induce cell wall extension in Plants. *Plant Cell* **1992**, *4*, 1425–1433. [CrossRef]
32. Ulvskov, P.; Wium, H.; Bruce, D.; Jørgensen, B.; Qvist, K.B.; Skjøt, M.; Hepworth, D.; Borkhardt, B.; Sørensen, S.O. Biophysical consequences of remodelling the neutral side chains of rhamnogalacturonan I in tubers of transgenic potatoes. *Planta* **2005**, *220*, 609–620. [CrossRef]
33. Lopez-Sanchez, P.; Martinez-Sanz, M.; Bonilla, M.R.; Wang, D.; Walsh, C.T.; Gilbert, E.P.; Stokes, J.R.; Gidley, M.J. Pectin impacts cellulose fibre architecture and hydrogel mechanics in the absence of calcium. *Carbohyd. Polym.* **2016**, *153*, 236–245. [CrossRef] [PubMed]

34. White, P.B.; Wang, T.; Park, Y.B.; Cosgrove, D.J.; Hong, M. Water–Polysaccharide Interactions in the Primary Cell Wall of *Arabidopsis thaliana* from Polarization Transfer Solid-State NMR. *J. Am. Chem. Soc.* **2014**, *136*, 10399–10409. [CrossRef] [PubMed]
35. Drozdov, A.D.; deClaville Christiansen, J. Modeling the effects of pH and ionic strength on swelling of anionic polyelectrolyte gels. *Model. Simul. Mater. Sci. Eng.* **2015**, *23*, 055005. [CrossRef]
36. Garnier, C.; Axelos, M.A.V.; Thibault, J.-F. Selectivity and cooperativity in the binding of calcium ions by pectins. *Carbohyd. Res.* **1994**, *256*, 71–81. [CrossRef]
37. Vicré, M.; Sherwin, H.W.; Driouich, A.; Jaffer, M.A.; Farrant, J.M. Cell wall characteristics and structure of hydrated and dry leaves of the resurrection plant *Craterostigma wilmsii*, a microscopical study. *J. Plant Physiol.* **1999**, *155*, 719–726. [CrossRef]
38. Moore, J.P.; Nguema-Ona, E.E.; Vicre´-Gibouin, M.; Sørensen, I.; Willats, W.G.T.; Driouich, A.; Farrant, J.M. Arabinose-rich polymers as an evolutionary strategy to plasticize resurrection plant cell walls against desiccation. *Planta* **2013**, *237*, 739–754. [CrossRef]
39. Leucci, M.R.; Lenucci, M.S.; Piro, G.; Dalessandro, G. Water stress and cell wall polysaccharides in the apical root zone of wheat cultivars varying in drought tolerance. *J. Plant Physiol.* **2008**, *165*, 1168–1180. [CrossRef]
40. Wakabayashi, K.; Hoson, T.; Kamisaka, S. Changes in amounts and molecular mass distribution of cell-wall polysaccharides of wheat (*Triticum aestivum* L.) coleoptiles under water stress. *J. Plant Physiol.* **1997**, *151*, 33–40. [CrossRef]
41. Wu, Y.; Spollen, W.G.; Sharp, R.E.; Hetherington, P.R.; Fry, S.C. Root growth maintenance at low water potentials (increased activity of xyloglucan endotransglycosylase and its possible regulation by abscisic acid). *Plant Physiol.* **1994**, *106*, 607–615. [CrossRef] [PubMed]
42. Iurlaro, A.; De Caroli, M.; Sabella, E.; De Pascali, M.; Rampino, P.; De Bellis, L.; Perrotta, C.; Dalessandro, G.; Piro, G.; Fry, S.C.; et al. Drought and heat differentially affect XTH expression and XET activity and action in 3-day-old seedlings of durum wheat cultivars with different stress susceptibility. *Front. Plant Sci.* **2016**, *7*, 1686. [CrossRef] [PubMed]
43. Choi, J.Y.; Seo, Y.S.; Kim, S.J.; Kim, W.T.; Shin, J.S. Constitutive expression of CaXTH3, a hot pepper xyloglucan endotransglucosylase/hydrolase, enhanced tolerance to salt and drought stresses without phenotypic defects in tomato plants (*Solanum lycopersicum* cv. Dotaerang). *Plant Cell Rep.* **2011**, *30*, 867–877. [CrossRef]
44. Chen, Y.; Zhang, B.; Li, C.; Lei, C.; Kong, C.; Yang, Y.; Gonget, M. A comprehensive expression analysis of the expansin gene family in potato (*Solanum tuberosum*) discloses stress-responsive expansin-like B genes for drought and heat tolerances. *PLoS ONE* **2019**, *14*, e0219837. [CrossRef] [PubMed]
45. Wu, Y.; Sharp, R.E.; Durachko, D.M.; Cosgrove, D.J. Growth maintenance of the maize primary root at low water potentials involves increases in cell wall extension properties, expansin activity and wall susceptibility to expansins. *Plant Physiol.* **1996**, *111*, 765–772. [CrossRef]
46. Wu, Y.; Thorne, E.T.; Sharp, R.E.; Cosgrove, D.J. Modification of expansin transcript levels in the maize primary root at low water potentials. *Plant Physiol.* **2001**, *126*, 1471–1479. [CrossRef]
47. Montechiarini, N.H.; Delgado, L.; Morandi, E.N.; Carrillo, N.J.; Gosparini, C.O. The expansin EXP1 gene in the elongation zone is induced during soybean embryonic axis germination and differentially expressed in response to ABA and PEG treatments. *Seed Sci. Res.* **2021**, *31*, 60–68. [CrossRef]
48. Zhou, P.; Zhu, Q.; Xu, J.; Huang, B. Cloning and characterization of a gene, *AsEXP1*, encoding expansin proteins inducible by heat stress and hormones in creeping bentgrass. *Crop Sci.* **2011**, *51*, 333–341. [CrossRef]
49. Michelena, V.A.; Boyer, J.S. Complete turgor maintenance at low water potentials in the elongating region of maize leaves. *Plant Physiol.* **1982**, *69*, 1145–1149. [CrossRef]
50. Pritchard, J.; Jones, R.W.; Tomos, A.D. Turgor, growth and rheological gradients of wheat roots following osmotic stress. *J. Exp. Bot.* **1991**, *42*, 1043–1049. [CrossRef]
51. Spollen, W.G.; Sharp, R.E. Spatial distribution of turgor and root growth at low water potentials. *Plant Physiol.* **1991**, *96*, 438–443. [CrossRef]
52. Wilkinson, S.; Davies, W.J. Xylem Sap pH increase: A drought signal received at the apoplastic face of the guard cell that involves the suppression of saturable abscisic acid uptake by the epidermal symplast. *Plant Physiol.* **1997**, *113*, 559–573. [CrossRef]
53. Mohnen, D. Pectin structure and biosynthesis. *Curr. Opin. Plant Biol.* **2008**, *11*, 266–277. [CrossRef] [PubMed]
54. Ishii, T.; Matsunaga, T.; Hayashi, N. Formation of rhamnogalacturonan II-borate dimer in pectin determines cell wall thickness of pumpkin tissue. *Plant Physiol.* **2001**, *126*, 1698–1705. [CrossRef] [PubMed]
55. Fleischer, A.; Titel, C.; Ehwald, R. The boron requirement and cell wall properties of growing and stationary suspension-cultured *Chenopodium album* L. cells. *Plant Physiol.* **1998**, *117*, 1401–1410. [CrossRef]
56. Sechet, J.; Htwe, S.; Urbanowicz, B.; Agyeman, A.; Feng, W.; Ishikawa, T.; Colomes, M.; Kumar, K.S.; Kawai-Yamada, M.; Dinneny, J.R.; et al. Suppression of Arabidopsis GGLT1 affects growth by reducing the L-galactose content and borate cross-linking of rhamnogalacturonan-II. *Plant J.* **2018**, *96*, 1036–1050. [CrossRef]
57. Zhao, C.; Zayed, O.; Zeng, F.; Liu, C.; Zhang, L.; Zhu, P.; Hsu, C.-C.; Tuncil, Y.E.; Tao, W.A.; Carpita, N.C.; et al. Arabinose biosynthesis is critical for salt stress tolerance in Arabidopsis. *New Phytol.* **2019**, *224*, 274–290. [CrossRef] [PubMed]
58. Byrt, C.S.; Munns, R.; Burton, R.A.; Gilliham, M.; Wege, S. Root cell wall solutions for crop plants in saline soils. *Plant Sci.* **2018**, *269*, 47–55. [CrossRef]

59. Rochange, S.F.; McQueen-Mason, S.J. Expression of a heterologous expansin in transgenic tomato plants. *Planta* **2000**, *211*, 583–586. [CrossRef] [PubMed]
60. Kiyosawa, K. Theoretical and experimental studies on freezing point depression and vapor pressure deficit as methods to measure osmotic pressure of aqueous polyethylene glycol and bovine serum albumin solutions. *Biophys. Chem.* **2003**, *104*, 171–188. [CrossRef]
61. Winzor, D.J. Reappraisal of disparities between osmolality estimates by freezing point depression and vapour pressure deficit methods. *Biophys. Chem.* **2004**, *107*, 317–323. [CrossRef] [PubMed]

plants

MDPI

Article

# With a Little Help from My Cell Wall: Structural Modifications in Pectin May Play a Role to Overcome Both Dehydration Stress and Fungal Pathogens

Ariana D. Forand [1], Y. Zou Finfrock [2,3], Miranda Lavier [3], Jarvis Stobbs [3], Li Qin [4], Sheng Wang [1,5], Chithra Karunakaran [3], Yangdou Wei [4], Supratim Ghosh [6] and Karen K. Tanino [1,*]

1 Department of Plant Sciences, University of Saskatchewan, Saskatoon, SK S7N 5A8, Canada; adf472@usask.ca (A.D.F.); sheng.wang@usask.ca (S.W.)
2 Advanced Photo Source, Lemont, IL 60439, USA; finfrock@anl.gov
3 Canadian Light Source, Saskatoon, SK S7N 2V3, Canada; Miranda.Lavier@lightsource.ca (M.L.); jarvis.stobbs@lightsource.ca (J.S.); chithra.karunakaran@lightsource.ca (C.K.)
4 Department of Biology, University of Saskatchewan, Saskatoon, SK S7N 5E2, Canada; li.qin@usask.ca (L.Q.); yangdou.wei@usask.ca (Y.W.)
5 Department of Biochemistry, Microbiology and Immunology, University of Saskatchewan, Saskatoon, SK S7N 5E5, Canada
6 Department of Food and Bioproduct Sciences, University of Saskatchewan, Saskatoon, SK S7N 5A8, Canada; supratim.ghosh@usask.ca
* Correspondence: karen.tanino@usask.ca

**Abstract:** Cell wall structural modifications through pectin cross-linkages between calcium ions and/or boric acid may be key to mitigating dehydration stress and fungal pathogens. Water loss was profiled in a pure pectin system and in vivo. While calcium and boron reduced water loss in pure pectin standards, the impact on *Allium* species was insignificant ($p > 0.05$). Nevertheless, synchrotron X-ray microscopy showed the localization of exogenously applied calcium to the apoplast in the epidermal cells of *Allium fistulosum*. Exogenous calcium application increased viscosity and resistance to shear force in *Allium fistulosum*, suggesting the formation of calcium cross-linkages ("egg-box" structures). Moreover, *Allium fistulosum* (freezing tolerant) was also more tolerant to dehydration stress compared to *Allium cepa* (freezing sensitive). Furthermore, the addition of boric acid ($H_3BO_3$) to pure pectin reduced water loss and increased viscosity, which indicates the formation of RG-II dimers. The *Arabidopsis* boron transport mutant, *bor1*, expressed greater water loss and, based on the lesion area of leaf tissue, a greater susceptibility to *Colletotrichum higginsianum* and *Botrytis cinerea*. While pectin modifications in the cell wall are likely not the sole solution to dehydration and biotic stress resistance, they appear to play an important role against multiple stresses.

**Keywords:** cell wall; pectin; calcium; boron; homogalacturonan; rhamnogalacturonan II; *Botrytis cinerea*; *Colletotrichum higginsianum*; dehydration; *Allium* and X-ray microscopy

check for updates

**Citation:** Forand, A.D.; Finfrock, Y.Z.; Lavier, M.; Stobbs, J.; Qin, L.; Wang, S.; Karunakaran, C.; Wei, Y.; Ghosh, S.; Tanino, K.K. With a Little Help from My Cell Wall: Structural Modifications in Pectin May Play a Role to Overcome Both Dehydration Stress and Fungal Pathogens. *Plants* **2022**, *11*, 385. https://doi.org/10.3390/plants11030385

Academic Editors: Penélope García-Angulo and Asier Largo-Gosens

Received: 14 December 2021
Accepted: 26 January 2022
Published: 30 January 2022

**Publisher's Note:** MDPI stays neutral with regard to jurisdictional claims in published maps and institutional affiliations.

**Copyright:** © 2022 by the authors. Licensee MDPI, Basel, Switzerland. This article is an open access article distributed under the terms and conditions of the Creative Commons Attribution (CC BY) license (https://creativecommons.org/licenses/by/4.0/).

## 1. Introduction

As the threat of climate change rages on, the frequency of dehydration stress and the risk posed by fungal pathogens is on the rise. These challenges are further complicated by the sessile nature of plants, which must survive a range of a combination of abiotic and biotic stresses. The cell wall serves as a barrier against stress, in addition to providing critical mechanical and structural support to the plant [1,2]. The ability of the cell wall to function in a range of mechanisms is dependent on the various components of the primary cell wall, including cellulose, hemicellulose, proteins and pectin [3]. Except for grasses, pectin is one of the most predominant components, comprising 30 to 50% of the cell wall dry matter [3]. Therefore, the composition of pectin is integral to the growth, morphology and development of the cell well, as well as in its ability to defend the plant against stress [4].

Pectin, the most structurally complex polysaccharide found in nature, is a five-member family currently considered to include: (1) Homogalacturonan (HG); (2) Rhamnogalacturonan I (RG-I); (3) Rhamnogalacturonan II (RG-II); (4) Xylogalacturonan (XGA); and (5) Apiogalacturonan (AP) [4]. HG and RG-II account for approximately 65% and 10% of all pectin, respectively, and have the unique ability to form more complex structures through cross-linkages to other elements [4–6].

Under the prevailing view, calcium ions form cross-linkages to carboxylate ions in demethylesterified galacturonic acid residues of HG [5,7]. The demethylesterifcation of HG occurs by way of pectin methylesterases (PMEs) [7] and pectin methylesterase inhibitors (PMEIs) provide negative feedback to PMEs [7]. In comparison, RG-II forms cross-linkages with boric acid, creating RG-II dimers [8]. These borate–diol ester cross link a boron atom and the apiosyl residue of side chain A in RG-II [8]. The formation of these cross-linkages maintains structural stability by reducing porosity while enhancing tensile strength and bound water in the cell wall fraction [9–16]. Increasing the quantity of bound water within the cell wall maintains tissue hydration and turgor pressure in addition to increasing cell wall rigidity [14]. Debra Mohnen's group recently proposed a new model for pectin structure consisting of a family of glycans (homoglycans, heteroglycans and proteoglycans) [17] and therefore, aspects of the current cell wall models may need to be reconsidered as more information emerges.

HG and RG-II have also been implicated as key forms of pectin in mitigating the resistance to desiccation and drought stress in several species, such as green algae and wheat cultivars [18–23]. This is likely the result of pectin forming hydrated gels, which in turn may limit damage during dehydration stress [19]. The degree of pectin methylation has also been linked to their ability to hold water, with reduced methylation increasing the water holding capacity of pectin [12]. Willats et al. (2001) also found the addition of calcium to pectin gels influences the water holding capacity, presumably as a result of "egg-box" structures [12].

The analysis of the *MUR1* gene has further implicated RG-II dimers as an important aspect of the tolerance to freezing stress, tied to dehydration stress resistance [24]. Despite these findings, there is a need for greater research regarding the role calcium and boron on the structure–function of the cell wall in relation to dehydration stress resistance.

Previous research has additionally highlighted the important role that pectin plays in allowing the cell wall to function as a barrier to *Botrytis cinerea*, a necrotrophic pathogen, and *Colletotrichum higginsianum*, a hemibiotrophic pathogen. Lionetti et al. (2007) observed increased resistance to *B. cinerea* in *Arabidopsis* as a result of the over-expression of PMEI1 and PMEI2 [25]. Other PMEIs have also been reported to increase plant resistance to *B. cinerea* [26]. Interestingly, PMEs play a beneficial role in the immunity against *B. cinerea*, despite the opposing nature of PMEIs and PME [27]. More generally, pectin has been identified as a main target during a *B. cinerea* infection [28]. Petrasch et al. (2019) reported that, during the process of a *B. cinerea* infection, the pathogen heavily degrades the cell wall and, in particular, pectin [28].

PMEIs also play a role in the immune response during a *C. higginsianum* infection [29]; however, research within this area is limited. A study conducted by Engelsdorf et al. (2017) found *Arabidopsis pmei6-2* mutants had reduced susceptibility to *C. higginsianum*. Reduced susceptibility may be indicative of a connection between the establishment of a *C. higginsianum* infection and pectin content [29]. Nonetheless, further exploration into the role of PMEIs and *C. higginsianum* is required.

Boron application has also been shown to play a positive role in *B. cinerea* infections in a variety of species [30,31]. Qin et al. (2010) found boron application reduced the germination of *B. cinerea* spores, reduced germ tube elongation and mycelial spread in table grapes [30]. When paired with boron, calcium increased resistance to *B. cinerea* in strawberry plants [31]. In addition, while there is no apparent research on the impact of boron on *C. higginsianum*, borate was found to inhibit the germination of *Colletotrichum gleosporioides* spores in mangos [32,33].

*Allium fistulosum* and *Allium cepa* were selected based on their contrasting freezing-induced dehydration stress tolerance. *A. fistulosum* is extremely tolerant to freezing stress, withstanding temperatures as low as −40 °C, while *A. cepa* lacks this ability [34,35]. Both species have easy to peel epidermal cell layers with a large cell size, making them an ideal species for this investigation. *Arabidopsis* was selected based on the wide range of genotypic mutant lines available and its close genetic relationship to *Brassica rapa*.

Since plants are often exposed to multiple stresses in the field, this study aims to gain a further understanding into the influence of calcium, boron and PMEI on pectin cross-linkages, and in turn, on the ability for the cell wall to act as a barrier against both water loss and fungal pathogens. We hypothesized the application of calcium and/or boron results in cell wall structural changes and increased resistance to both abiotic and biotic stress in *Allium* species and *Arabidopsis thaliana*.

## 2. Method and Materials

### 2.1. Plant Material and Growth Conditions

*A. fistulosum* seeds and Yellow Sweet Spanish *A. cepa* seeds (Early's Farm & Garden Center, Saskatoon, SK, Canada) were grown in the Agriculture Greenhouses, University of Saskatchewan (Saskatoon, SK, Canada) in 6″ pots containing Sunshine Mix No. 4 (Sungro Horticulture Canada Ltd. Seba Beach, AB, Canada) at approximately 25/22 °C (day/night) under natural light supplemented with high pressure sodium lights (17 h photoperiod, average 600 μmol $m^{-2}s^{-1}$). Watering (City of Saskatoon, SK, Canada, water) was applied every second day during the spring/summer months and every third day during the fall and winter months with 200 mL of 20–20–20 (NPK) fertilizer (150 g $L^{-1}$) (Plant Prod, Brampton, ON, Canada) weekly during the spring/summer months and twice weekly during the fall/winter months. The experiment was arranged in a randomized complete block design across the bench.

In addition, five *A. thaliana* genotypes were analyzed: three boron transporter mutants (*nip5;1-1* (SALK_122287C), *nip6;1-2* (SALK_046323C) and *bor1-3* (SALK_037312)), a pectin methylesterase inhibitor over-expression mutant (*p35S::PMEI5*) and a wild-type (Col-0) line, respectively [36–39]. The three boron mutants were obtained from the *Arabidopsis* Biological Resource Centre (ABRC) (Columbus, OH, USA). The transgenic line overexpressing PMEI5 driven by the Cauliflower mosaic virus (CaMV) 35S promoter was provided courtesy of Kerstin Müller [39]. Lines were genotyped according to Edwards et al. (1991), with some modifications [40,41]. Gene specific primers were ordered from Integrated DNA Technologies (Coralville, IA, USA) (Table S1).

*Arabidopsis* plants were grown in the Agriculture phytotron (Conviron, Winnipeg, MB, Canada) under 20 °C constant temperature, 50% RH, 16 h photoperiod, with an irradiance of 150 ± 10 $m^{-2}s^{-1}$, and watered every second day (City of Saskatoon, SK, Canada). Two g/L fertilizer (20–20–20 NPK) was applied weekly.

### 2.2. Forms of Calcium and Boron Application

Calcium treatments of 100 mL of a 0.05 M $CaCl_2$ (Fisher Scientific, Watham, MA, USA) were applied every second day for four weeks as a soil drench. For pectin solutions prepared with calcium or boron, 0.05 M of $CaCl_2$ or $H_3BO_3$ were added directly during the preparation of the pectin solution to ensure homogeneity.

### 2.3. Rheology

Two types of pectin were examined, TIC Pretested Pectin HM (69–75% methylation) [42] slow set (standardized with dextrose) (TIC Gums, White Marsh, MA, USA) and GENU BETA pectin (55% methylation) [43] (standardized with EU non-GM beet sucrose) (CP Kelco Atlanta, GA, USA). The HM pectin is derived from citrus pectin and as a result has a high HG content and a lower RG content, making it similar to the pectin found within the cell wall of *Allium* spp. [44,45]. GENU BETA pectin is produced from sugar beets, and as such it has a higher RG-II content (~5% compared to 0.6%), making it more like the pectin found within *Arabidopsis*. Therefore, HG pectin was utilized as a proxy for *Allium*, while GENU BETA pectin was selected as a proxy for *Arabidopsis*.

Pectin viscosity was measured using the AR G2 rheometer (TA Instruments, New Castle, DE, USA) with the 40 mm smooth acrylic geometry. Measurements were taken at 20 °C, 12 °C and 4 °C. Viscosity was also measured throughout the temperature ramp. Prior to the first set of measurements at 20 °C, the sample was conditioned to 20 °C with a soak time of 120.0 $s^{-1}$ and a duration of 60.0 $s^{-1}$. Viscosity measurements were then taken for 300.0 $s^{-1}$ with a shear rate of 0.11 $s^{-1}$. The sampling rate for the measurements taken was 10 $s^{-1}$/pt. The temperature was then decreased to 12 °C at a rate of 5.0 °C/min, with the shear rate remaining constant at 0.11 $s^{-1}$. The sample was then conditioned with the same soak time and duration as stated above. Viscosity measurements were also conducted under the same conditions previously outlined. The temperature was decreased to 4 °C and measurements of viscosity were performed as previously described above. The experiment was repeated three independent times with four replicates of each treatment group per trial ($n$ = 12).

### 2.4. Texture Analysis

The force required to shear *A. fistulosum* sheaths was evaluated using the 10-blade Allo–Kramer shearing compression cell, attached to the TMS-Pro Texture Press (Food Technology Corp., Sterling, VA, USA) in the Shand lab at the University of Saskatchewan. Calcium-treated plants received 100 mL soil drench of a 0.05 M $CaCl_2$ solution every second day for four weeks. Each measurement was performed using three 4 cm sections of *A. fistulosum* sheaths, from the youngest leaf with the most developed sheath. The ligule and the epidermal cell layer were left intact. The full-scale load was 1000 N, the crosshead speed was 500 mm $min^{-1}$. The shear force (in Newtons) required to shear 1 g of fresh sample was calculated using Equation (1) [46]. The shear force was measured in N $g^{-1}$ [46]. The experiment was repeated three independent times, with four replicates per trial ($n$ = 12).

$$Force\ Required\ to\ Shear\ Allium\ Sheaths = \frac{peak\ shear\ force\ (\mathrm{N})}{weight\ of\ sheaths\ (\mathrm{g})} \tag{1}$$

Equation (1). Force required to shear *Allium fistulosum* sheaths in N $g^{-1}$.

### 2.5. Water Loss in Pure Pectin Standards

TIC Pretested Pectin HM (69–75% methylation) slow set (standardized with dextrose) was used as a model for pectin within *Allium* and GENU BETA pectin (55% methylation) (standardized with EU non-GM beet sucrose) was a model for *Arabidopsis* pectin. Treatments consisted of pectin solutions at concentrations of 4% and 8%, with and without 0.05 M of $CaCl_2$ or 0.05 M of $H_3BO_3$. Overall, 12 different pectin solutions were examined with 4 replications per pectin solution treatment, and the experiment was repeated 3 times.

Gravimetric water loss on an analytical balance was recorded at hourly intervals over 0–6 h in the various pectin solutions. This timeline was selected following preliminary experiments. Approximately 1 g of each of the pectin solutions was evenly pipetted into Petri dishes from the center point. Throughout the course of the dehydration experiment, the lids were kept off the Petri dishes to allow for evaporation. The temperature remained constant at 23 °C, with a relative humidity of ~22%. The percentage water loss was measured as per Equation (2).

$$Percentage\ Water\ Loss = (1 - \left(\frac{T0\ weight}{T(x)weight}\right) \times 100) \tag{2}$$

Equation (2). Percentage water loss in pure pectin standards. Variable "*T*" is equal to the mass of the plant tissue at a specified time point.

### 2.6. Allium Dehydration Stress Tolerance

Dehydration tolerance was measured by first analyzing the percentage water loss over 16–18 h using 4 cm sections of *A. fistulosum* and *A. cepa* sheaths obtained from the youngest leaf with the most developed sheath. Both ends of the sheath were sealed with Vaseline® and weighed to obtain a T0 weight. The sheaths were then dehydrated under dark conditions at 23 °C and ~33% RH. Weights were recorded at 16 h and 18 h. Following dehydration, the sheaths were wrapped in moist Kim Wipes and placed in 50 mL Falcon tubes with the lid for 24 h. The percentage water loss was measured and calculated (Equation (2)).

Following dehydration, viability was assessed based on the presence/absence of protoplasmic streaming and staining using fluorescein diacetate (FDA). Epidermal cell layers were peeled from the sheath prior to viability analysis. Protoplasmic streaming was observed using a digital LEICA DM4 B microscope (Wetzlar, Germany) at 40× with a LEICA DFC7000 T camera (Wetzlar, Germany). Protoplasmic streaming was quantified as a percentage of viability across the epidermal cell layer (Equation (3)). The same microscope was used to visualize epidermal cell layers at 40× following FDA staining.

$$Percent\ Cell\ Viability = \frac{number\ of\ cells\ with\ protoplasmic\ streaming}{total\ number\ of\ cells} \tag{3}$$

Equation (3). Percent cell viability based on protoplasmic streaming. Total cell count was measured by counting the number of cells within the frame at the 40× objective.

### 2.7. Arabidopsis Dehydration Stress Tolerance

Tolerance to dehydration stress was assessed by measuring percentage water loss every hour over 10 h in the total above ground biomass of the various two-week-old genotypes. An initial weight was recorded at 0 h and then every 2 h until 10 h. The percentage water loss was measured and calculated (Equation (2)). The samples were kept under dark conditions at 23 °C and ~33% RH. Prior to dehydration, the severed end of the shoot was sealed using Vaseline®. Following dehydration, plants were rehydrated in 2 mL sepia toned bottles with 100 µL of $dH_2O$.

Viability following dehydration was assessed using electrical conductivity (Twin Compact Meter, Horiba, Japan) by first adding 1000 µL of $dH_2O$ to the bottles ~24 h following rehydration and then placing the bottles on the shaker for ~19 h at 23 °C. A second total electrical conductivity measurement was then taken after placement in a 100 °C dry bath for 10 min and then vortexed. Percentage electrolyte leakage (µS/cm) was measured and calculated (Equation (4)) [47].

$$Perecent\ Electrolyte\ Leakage = \frac{\left(\left(\frac{Pre-Boil\ Conductivity\ Value}{Post-Boil\ Conductivity\ Value}\right) \times 100\right)}{T0\ weight} \tag{4}$$

Equation (4). Percent electrolyte leakage in *Arabidopsis thaliana* above ground biomass following dehydration, where *T*0 was the time at 0 h.

The experiment was conducted under dark conditions to minimize the amount of water lost through open stomata. Stomatal closure was confirmed using the Suzuki Universal Micro-Printing (SUMP) method [48]. SUMP discs and SUMP liquid were used to take imprints of stomata, while a digital LEICA DM4 B microscope (Wetzlar, Germany) at 40× with a LEICA DFC7000 T camera (Wetzlar, Germany) attached assessed the imprints.

### 2.8. Calcium Localization

Calcium was spatially localized within single epidermal cell layers obtained from *A. fistulosum* using X-ray microscopy. Plants were grown and treated with calcium chloride as outlined above. The peeled epidermal layers were laid flat on Kapton tape and were used for data collection at the Advanced Photon Source (APS) beamline (20-ID:

Sector 20—Insertion Device Beamline) (Lemont, IL, USA). Characteristics of the X-ray beam: incident energy—10 keV; sample scanning area—160 μm × 160 μm; scanning step size—1 μm; and dwell time—10 milliseconds per pixel. A total of 10 vertical scan maps were collected at a depth of every 2 μm into the sample [49].

Two-dimensional images were processed using PyMCA to fit the normalized average spectra of all points on the map and to generate the calcium distribution map (Version 5.3.1) [50]. OriginPro (Northhampton, MA, USA) was then used to plot the calcium map and to increase image quality by eliminating the saturated pixels. False color maps were produced using rainbow color scheme in which the blue-to-red color gradient shows low-to-high relative calcium concentrations.

The completed images were then analyzed using the "histogram" feature on ImageJ (Version 1.53a) to gain a greater understanding into differences in RGB pixels between control and calcium-treated epidermal cell layers.

### 2.9. *Colletotrichum higginsianum* Infection

*C. higginsianum* Sacc culture, isolate IMI349061, originating from *Brassica rapa* (CABI Bioscience), was prepared as outlined in Liu et al. (2010) and Liu et al. (2007) [51,52]. Spores were suspended in dd$H_2O$, with a final concentration of $10^6$ spores/mL. Plants were inoculated by applying multiple single 1 μL droplets of inoculant to the oldest leaf, avoiding the mid-vein.

Throughout the course of infection, leaves were maintained on moist filter paper in Petri dishes kept under a 16–8 h/light–dark period (150 ± 10 μmol $m^{-2}s^{-1}$) during the first 2 and last 2 days of the period of infection, at the third day the plates were transferred to complete darkness. Complete darkness was used to promote germination of the appressorium, while the leaves were transferred back to a day/night light cycle on days 4–5 to prevent chlorosis at approximately 20 °C throughout.

Infection was then evaluated every 24 h based on the percent total leaf area infected (Equation (5)). The lesion area was measured using ImageJ (Version 1.53a).

$$Percent\ leaf\ area\ infected = \left(\frac{Lesion\ area\ (\mathrm{cm}^2)\ at\ T(x)}{Total\ leaf\ area\ (\mathrm{cm}^2)\ at\ T(x)}\right) \times 100 \tag{5}$$

Equation (5). Percent leaf area infected by *Botrytis cinerea* and *Colletotrichum higginsianum*, where $T(x)$ was the time point of interest.

### 2.10. *Botrytis cinerea* Infection

*B. cinerea* grown on potato dextrose agar (PDA) for 7 days at 22 °C was used to inoculate four-week-old *A. thaliana* leaves [53]. The leaves were inoculated using a solution containing *B. cinerea* mycelium suspended in dd$H_2O$, with a final concentration of $10^6$ spores/mL. Three 1 μL droplets of the mycelium containing solution were applied to one half of the leaf, while three 1 μL droplets of water were applied to the other half. Caution was taken to avoid the leaf mid-vein. Throughout the course of infection (120 h), the leaves were maintained on moist filter paper within sealed Petri dishes and kept the dark at 23 °C throughout the course of infection.

The infection was evaluated based on the percent total leaf area infected (Equation (5)).

### 2.11. Statistical Analysis

Analyses related to shear force, water loss, protoplasmic streaming, electrolyte leakage, and fungal pathogens were performed using ANOVA tests, with Tukey's tests as a method of post hoc analysis ($p < 0.05$). Two-tailed two-sample *t*-tests ($p < 0.05$) analyzed calcium localization according to differences in RGB pigmentation and differences in the intensity of green pixels following FDA staining.

Rheological results were analyzed using generalized additive models (GAMs). Models were initially fit based on each fixed explanatory variable separately, prior to fitting models at all fixed explanatory variables together. Data were log-transformed. GAMs were generated using the gam function from the "mgcv" package [54,55]. The Akaike information criterion (AIC) score was used when building models to select for the optimum model. Function gam.check was used for model checking. GAMs were analyzed using ANOVA, with F-tests. Figures of each GAM were individually created for each of the pectin solutions using function "plot.gam", in addition to the packages "mgcViz" and "rgl" [56,57]. The package "devtools" was used to download the color palette, and "inauguration_2021" was used for coloration figures [58].

All statistical analyses were performed with the RStudio statistical software (Version 1.2.5033). In addition to the previous packages mentioned, "ggplot" and "ggplot2" were also used [58].

## 3. Results

Pectin viscosity significantly ($p < 0.05$) increased under the application of boron alone in both HM pectin and GENU BETA sugar beet (GB) pectin (Figure S1, Table S2). Boron had a more pronounced impact on 8% HM pectin viscosity at 5 °C, increasing the viscosity of HM pectin to just over 6000 Pa.s 5 °C compared to 8% GB pectin with boron, which had an average viscosity of less than 4000 Pa.s at 5 °C (Figure 1). Boron also had a more notable impact on HM pectin at 12 °C, with the trend only changing at 20 °C, when boron increased GB pectin viscosity to levels greater than HM pectin viscosity (Figure 1). Interestingly, calcium application alone did not have a significant effect on pectin viscosity (Table S2). As outlined in Section 2.2, calcium and boron were added directly to the pectin solution. However, calcium did have a significant effect on pectin viscosity when its influence was considered in combination with increasing pectin concentration and temperature (Table S2). The effect of boron on pectin viscosity remained significant when its effect was considered in combination with increasing pectin concentration and temperature (Table S2).

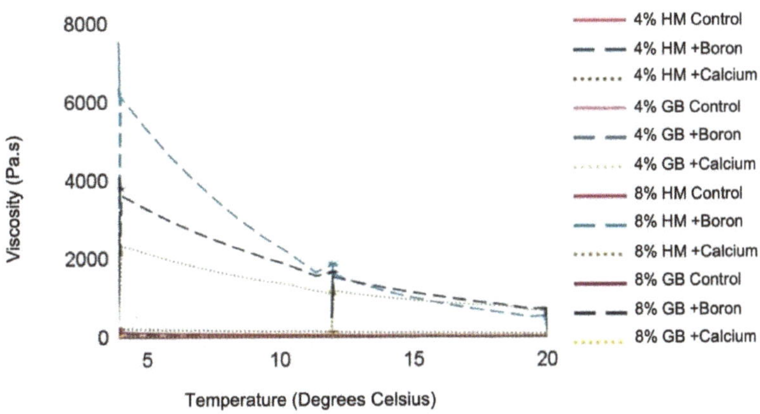


**Figure 1.** Change in viscosity (Pa.s) as temperature (°C) increased. Viscosity was analyzed across two concentrations of pectin (4% or 8%), two types of pectin (high methylated citrus pectin or GENU BETA sugar beet pectin) and with either calcium (0.05 M $CaCl_2$), boron (0.05 M $H_3BO_3$) or no additional element.

The analysis of the force required to shear *A. fistulosum* sheaths revealed the application of calcium (applied as a soil drench) resulted in a significantly ($p < 0.05$) higher Allo–Kramer shear force (265.0 N $g^{-1}$), compared to *A. fistulosum* plants that had not received calcium (210.6 N $g^{-1}$) (Figure 2). This increase in shear force is representative of an increased toughness within the sheaths. Toughness is a mechanical property that describes the ability for a material to resist fracture [59,60]. The toughness of a material is influenced by both its strength and ductility [59,60].

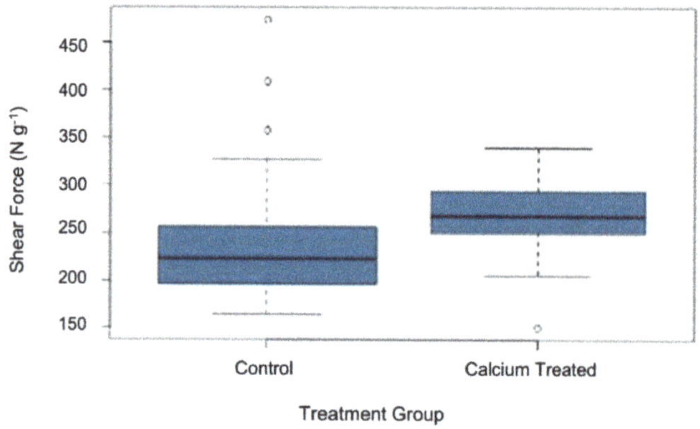


**Figure 2.** The effect of calcium application on the force (N $g^{-1}$) required to shear through *Allium fistulosum* sheaths. Plants treated with calcium received 100 mL of a 0.05 M $CaCl_2$ solution every second day. See Table S3 for statistical analysis.

Calcium was spatially localized within *A. fistulosum* epidermal cell walls (Figure 3). X-ray microscopy imaging showed that the exogenous application of calcium to *A. fistulosum* increased the concentration of calcium within the cell layer, with the additional calcium primarily localizing to the apoplast (Figure 3). The spatial localization of calcium to the apoplast is evident in Figure 3B, where epidermal layer cell walls can be observed because of increased calcium concentrations in the apoplast (more intense lighter blue color). This calcium appears to be primarily distributed to the radial cell walls. These areas of interest are identified using white arrows. Small flecks of white, light blue and green are also visible across both Figure 3A,B. These, in addition to red flecks, which are more pronounced on Figure 3B, represent calcium contamination and are not related to structures within the cell layer.

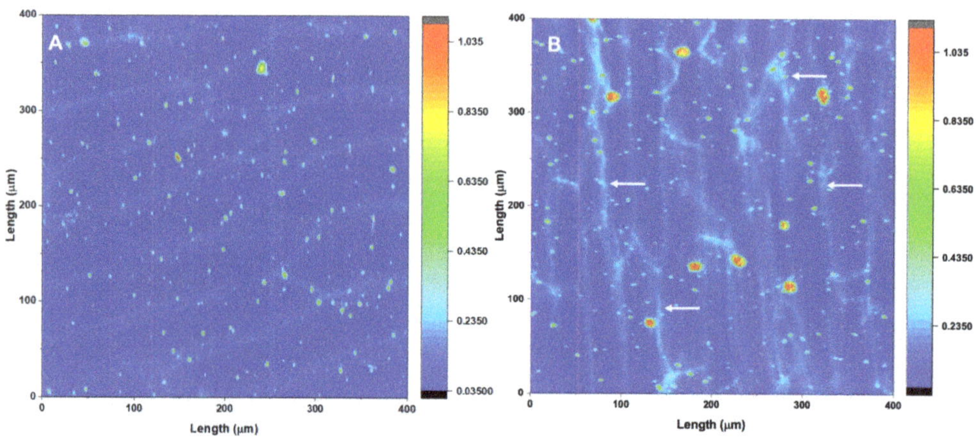


**Figure 3.** 2D Calcium map obtained using the 20-ID beamline at the Advanced Photon Source, Lemont, IL, and the X-ray microscopy technique. Calcium concentration is represented by color changes within the map, with darker blue indicating a lower concentration of calcium to red indicating a higher concentration of calcium. White arrows point to sample areas where calcium appears to localize to the apoplast. (**A**) Single *Allium fistulosum* epidermal cell layer obtained from a control plant without calcium treatment. (**B**) Single *Allium fistulosum* epidermal cell layer obtained from a plant treated with a soil drench of 100 mL of a 0.05 M $CaCl_2$ every second day for four weeks.

In addition to increasing the toughness of *A. fistulosum* sheaths, calcium application significantly ($p < 0.05$) reduced percentage water loss, as did increasing pectin concentrations (Figure 4B). An 8 percent high methylated citrus pectin with calcium (proxy for *Allium*) lost the least amount of water after 6 h (48.5%), while 8% GENU BETA pectin with calcium (proxy for *Arabidopsis*) had a water loss of 58.3% after 6 h (Figure 4A). However, the effect of calcium in mitigating dehydration stress in *Allium* species tissue was not significant ($p > 0.05$) (Table S4).

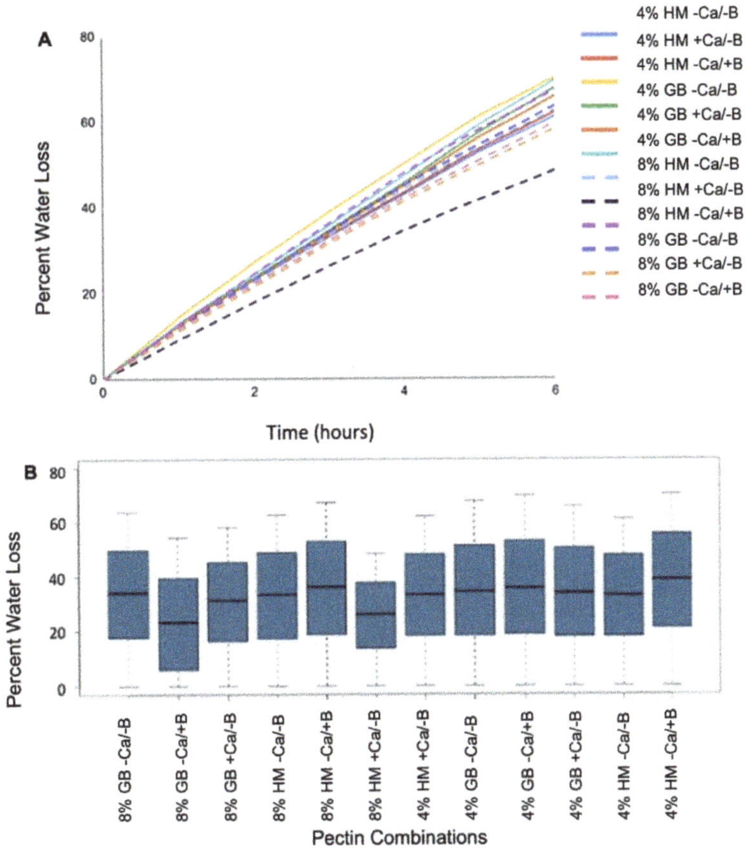


**Figure 4.** (**A**) Percent water loss over 6 h over a 6 h period for the 12 pectin solutions of interest. Amongst the 12 solutions, there are two concentrations (4% or 8%), two pectin types (high methylated (HM) citrus pectin or GENU BETA (GB) (sugar beet) pectin), and the addition of calcium (Ca) (0.05 M $CaCl_2$), boron (B) (0.05 M $H_3BO_3$) or no additional element. (**B**) Boxplot of average percent water loss over a 6 h period for the 12 pectin solutions of interest. Amongst the 12 solutions, there are two concentrations (4% or 8%), two pectin types (high methylated (HM) citrus pectin or GENUA BETA (GB) (sugar beet) pectin), and the addition of calcium (Ca) (0.05 M $CaCl_2$), boron (B) (0.05 M $H_3BO_3$) or no additional element. See Tables S5 and S6 for statistical analysis.

The freezing-tolerant *A. fistulosum* was also significantly ($p < 0.05$) more resistant to dehydration stress compared to the freezing-sensitive *A. cepa* (Figure 5B). After 17 h, *A. fistulosum* sheaths had a percentage water loss of 27.1%, while sheaths obtained from *A. cepa* lost 33.1% water (Figure 5A). *A. fistulosum* continued to lose less water as the period of dehydration was extended from 16 h to 18 h, retaining 5.2% more water compared to *A. cepa* (28.1% compared to 33.3%) (Figure 5A).

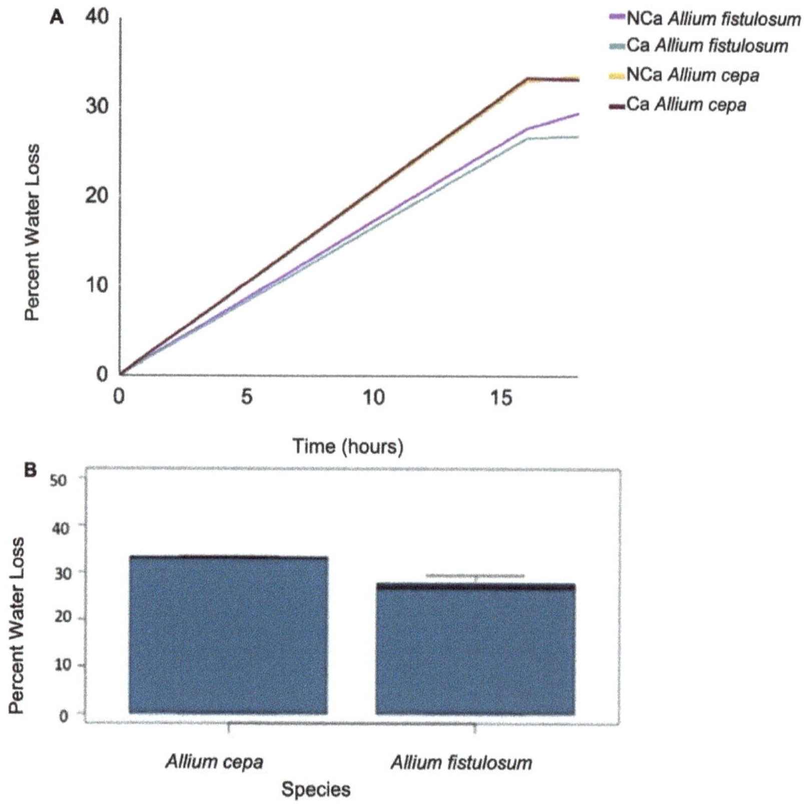


**Figure 5.** (**A**) Percent water loss over 16–18 h in *Allium fistulosum* and *Allium cepa*. Plants treated with calcium (CA) received 100 mL of a 0.05 M $CaCl_2$ every second day for four weeks. Control plants (NCA) did not receive calcium. (**B**) Boxplot showing average percent water loss over 16–18 h in *Allium fistulosum* and *Allium cepa* sheaths. Each bar represents both control and calcium treated plants combined. See Table S4 for statistical analysis.

In addition, epidermal cell layers from dehydrated and subsequently rehydrated *A. fistulosum* sheaths had a significantly ($p < 0.05$) higher percentage viability based on protoplasmic streaming compared to *A. cepa* (Figure 6B). After 16 h of dehydration, cell layers obtained from *A. fistulosum* had a 57.6% viability compared to 7.1% in *A. cepa* (Figure 6A). The trend of increased viability in *A. fistulosum* continued as the length of dehydration was extended. After 18 h, *A. fistulosum* epidermal cell layers had 39.2% viability while those obtained from *A. cepa* had a percent viability of 2.5% (Figure 6A).

In addition, there was a higher level of FDA-based "greenness" within calcium treated cell layers, indicating an increased level of viability (Figures S2 and S3). Generally, cell layers obtained from *A. fistulosum* also had a significantly ($p < 0.05$) increased level of greenness compared to those obtained from *A. cepa*, indicating greater viability following dehydration stress (Table S8). In addition, differences amongst other relationships were also found to be statistically significant ($p < 0.05$) (Table S8).

Boron also significantly ($p < 0.05$) reduced water loss in pure pectin solutions (Figure 4B). After 6 h, 8% GENU BETA pectin with boron (proxy for *Arabidopsis*) had a 59.7% water loss compared to 63.7% for the GENU BETA pectin control (Figure 4A). While similar dehydration stress tolerance trends were observed in *Arabidopsis* boron transporter mutants, the differences were non-significant ($p > 0.05$) (Figure 7B). Moreover, the over-expression of *PMEI5* did not significantly influence dehydration stress resistance ($p > 0.05$) (Figure 7B).

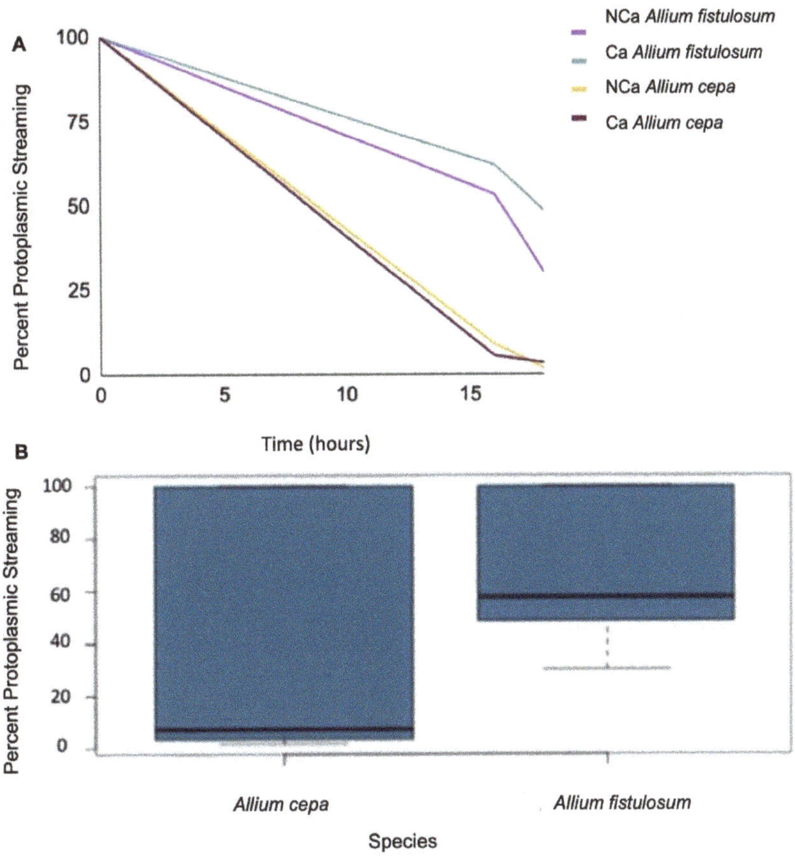


**Figure 6.** (**A**) Percent protoplasmic streaming in *Allium fistulosum* and *Allium cepa* epidermal cell layers following dehydration over 16–18 h. Plants treated with calcium (CA) received 100 mL of a 0.05 M $CaCl_2$ every second day for four weeks. Control plants (NCA) did not receive calcium. (**B**) Boxplot showing percent protoplasmic streaming for *Allium fistulosum* and *Allium cepa*, following a 16–18 h dehydration period. Each bar represents both control and calcium treated plants combined. See Table S7 for statistical analysis.

However, mutations in boron transporters and the over-expression of *PMEI5* were found to influence the rate of infection for both fungal pathogens (Figure 7). The analysis of the rate of *B. cinerea* infection amongst the genotypes of interest found *bor1* to be the most susceptible to the pathogen, consistently having the greatest percent area of infection from 1 dpi (days post-inoculation) (Figure 7). In general, the percentage of leaf area infected in the *bor1* leaves was significantly ($p < 0.05$) greater compared to all the other genotypes of interest (Figure 7B). At 5 dpi, 100% of the *bor1* leaf tissue was covered in a *B. cinerea* lesion (Figure 7A). By comparison, *nip6;1* had 60.7% of its leaf area infected, leaves obtained from *nip5;1* were 61.4% covered in a *B. cinerea* lesion, while *p35S::PMEI5* leaves were 68.4% infected (Figure 7A). Col-0 leaves were 71.4% covered by the lesion (Figure 7A).

In general, over the entire course of the infection, *bor1* was also significantly ($p < 0.05$) more susceptible to *C. higginsianum* compared to Col-0 (Figure 8B). At 5 dpi, over 60% of *bor1* leaf tissue was covered in a lesion caused by *C. higginsianum*, while all other genotypes of interest had less than 40% of their leaf tissue covered by a lesion (Figure 8A). At 1 dpi, the percentage leaf area covered by lesions from *C. higginsianum* was over 30% larger compared to the other genotypes analyzed (Figure 8A).

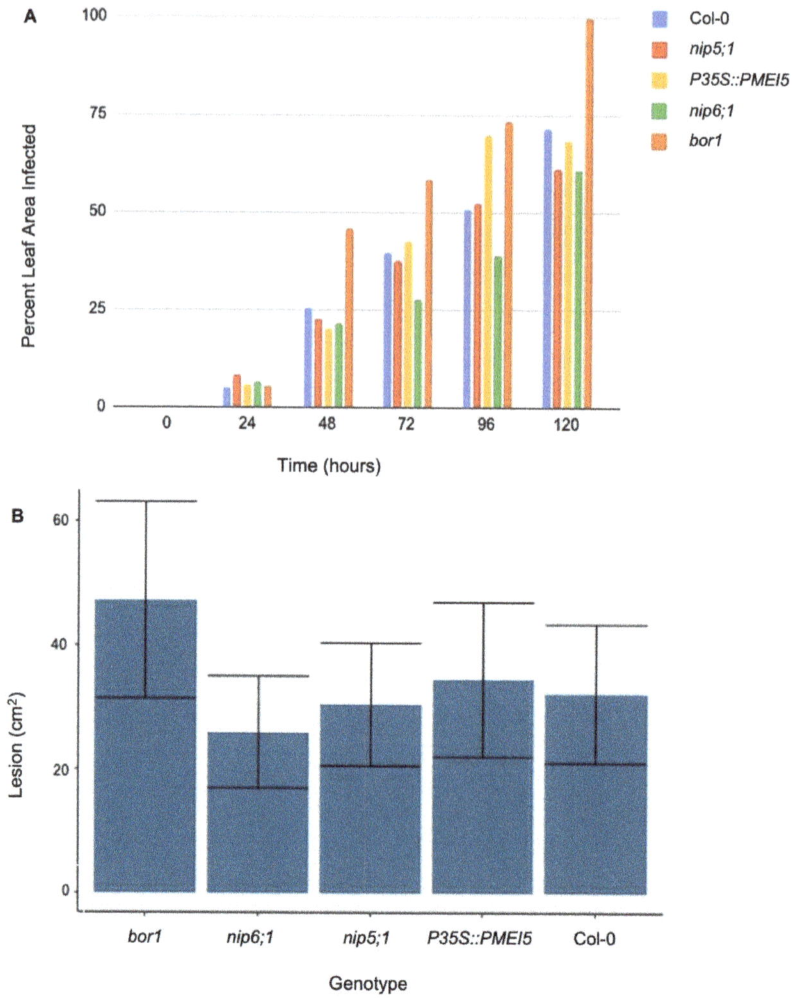


**Figure 7.** (**A**) Average lesion size on *Arabidopsis thaliana* leaves caused by *Botrytis cinerea*. Five genotypes were examined (*bor1* (boron transporter mutant), *nip5;1* (boron transporter mutant), *nip6;1* (boron transporter mutant), *p35S::PMEI5* (*PMEI* overexpression), and Col-0 (wild type)) over a 0–120 h period post-inoculation. (**B**) Boxplot showing average lesion size on *Arabidopsis thaliana* leaves caused by *Botrytis cinerea*. Five genotypes were examined (*bor1* (boron transporter mutant), *nip5;1* (boron transporter mutant), *bor1* (boron transporter mutant), *nip6;1* (boron transporter mutant), *p35S::PMEI5* (PMEI mutant), and Col-0 (wild type)) over a 0–120 h period post-inoculation. Error bars represent standard error. See Tables S9 and S10 for statistical analyses.

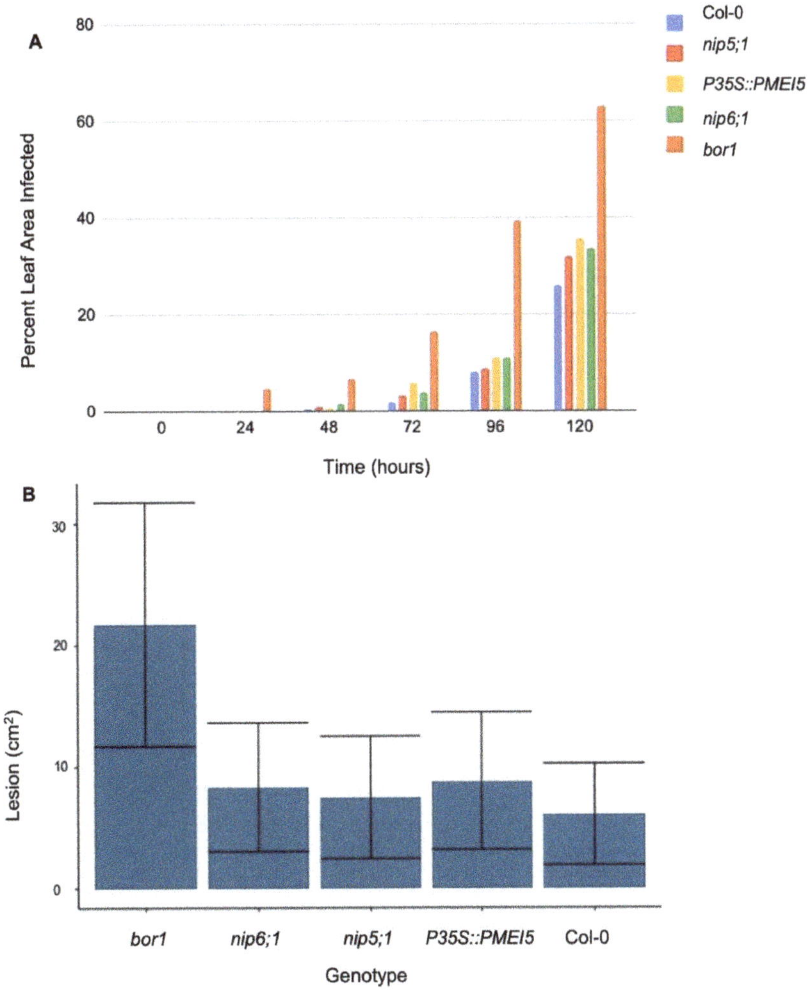


**Figure 8.** (**A**) Average lesion size on *Arabidopsis thaliana* leaves caused by *Colletotrichum higginsianum*. Five genotypes were examined (*bor1* (boron transporter mutant), *nip5;1* (boron transporter mutant), *bor1* (boron transporter mutant), *nip6;1* (boron transporter mutant), *p35S::PMEI5* (PMEI mutant), and Col-0 (wild type)) over a 0–120 h period post-inoculation. (**B**) Boxplot showing average lesion size on *Arabidopsis thaliana* leaves caused by *Botrytis cinerea*. Five genotypes were examined (*bor1* (boron transporter mutant), *nip5;1* (boron transporter mutant), *bor1* (boron transporter mutant), *nip6;1* (boron transporter mutant), *p35S::PMEI5* (PMEI mutant), and Col-0 (wild type)) over a 0–120 h period post-inoculation. Error bars represent standard error. See Tables S11 and S12 for statistical analyses.

## 4. Discussion

### 4.1. Mechanical Changes and Calcium Localization

The structural role that pectin plays in the cell wall is critical to a plant's ability to withstand stress [4]. Within the cell wall, pectin is far from the only critical component. However, what sets pectin apart from other components of the cell wall is the unique ability for HG and RG-II to form complex structures through cross-linkages to other elements [5,7]. Both the rheological results and those obtained from the analysis of shear force are suggestive that structural changes occurred because of calcium and boron application. Furthermore, the addition of these elements likely resulted in the formation of "egg-box" structures and RG-II dimers, as they are known to influence cell wall integrity, including rigidity and tensile strength [6,7,61].

The notion that calcium–HG cross-links form within *A. fistulosum* epidermal cell walls because of calcium application is further supported by images obtained from the Advanced Photon Source (Lemont, IL, USA) using the 20-ID beamline, the X-ray microprobe technique and X-ray fluorescence. These images clearly show an accumulation of calcium around the cell walls of epidermal cells obtained from calcium treated *A. fistulosum* and are a key piece of the puzzle in the narrative of the formation of "egg-box" structures. Within plant cells, pectin is only found within the cell wall and is most highly concentrated in the middle lamella and the primary cell wall [4]. Thus, the localization of calcium to the cell wall/middle lamella region and the increased shear force suggest that the application of calcium to *A. fistulosum* results in the formation of calcium–HG cross-links. Our next step was to analyze how the likely formation of these structures, in addition to boron and PMEIs, would influence dehydration stress resistance.

### 4.2. Dehydration Stress in Allium spp. and Pectin Proxy

In addition to causing structural changes within plant tissue, the formation of calcium–HG cross-linkages has also been tied to reduced cell wall permeability [12]. Thus, it was hypothesized the application of calcium would increase resistance to dehydration stress tolerance by reducing percentage water loss.

Even though calcium was found to significantly ($p < 0.05$) reduce percentage water loss in high methylated citrus pectin (proxy for *Allium*), the role calcium played on improving dehydration stress resistance in *A. fistulosum* and *A. cepa* was non-significant ($p > 0.05$) and is indicative of the complex nature of plants compared to pure systems. Nevertheless, previous studies have observed the efficacy of calcium application in enhancing drought stress tolerance/resistance in species such as *Beta vulgaris*, *Nicotiana tabacum* and wheat [62–64].

The discrepancy between these previous findings and those within this paper may be the result of a multitude of reasons, one of them being pore size. While the exact diameter of pores within the *A. fistulosum* epidermal cell layer is not known, previous research has found that, under cold acclimation, the diameter is reduced to less than 1.3 nm; however, the diameter of a water molecule at 25 °C is ~0.27 nm (2.7 Å) [65,66]. Therefore, even if permeability was reduced as a result of calcium application, the pore size was still larger than that of a water molecule. The large volume of water within the sheaths of *A. fistulosum* and *A. cepa* may also mask the effect of the calcium treatment.

Calcium is a complex element within plant systems, playing a role in a wide range of functions, including acting as a signaling molecule [67]. In its role as a signaling molecule, calcium has also been found to influence tolerance and resistance to drought stress in a variety of species [63,64,68]. For example, Knight et al. (1997) observed changes in the concentration of free calcium within cytosol during drought stress treatment in Arabidopsis [68]. More generally, the three main calcium sensors (1) calcineurin B-like protein; (2) calmodulin, and calmodulin-like proteins; and (3) calcium-dependent protein kinases) transduce $Ca^{2+}$ signals, in turn causing the phosphorylation of a downstream target [69–76]. This ultimately results in a response to drought stress.

While calcium was unable ($p > 0.05$) to improve the overall resistance to dehydration in either *A. fistulosum* or *A. cepa*, *A. fistulosum* was significantly ($p < 0.05$) more tolerant to dehydration stress. *A. fistulosum* has also been found to be more tolerant to freezing-induced dehydration as compared with *A. cepa* [34,35]. *A. fistulosum* tolerates temperatures around −40 °C, while *A. cepa* lacks the ability to cold acclimate beyond −11 °C [34,35]. The connection between dehydration stress resistance and freezing stress tolerance is critical in the face of climate change. While initially drought stress and freezing stress may seem unrelated, both are tied to dehydration stress. During both drought stress and freezing stress, loss of turgor pressure can occur as the cell loses water [77–79]. As a result, cytorrhysis may occur beyond a tolerable limit. Oertli (1986) and Oertli et al. (1990) reported that the ability for the cell wall to resist cytorrhysis depends on the mechanical strength of the cell wall [80,81]. Therefore, the ability to resist dehydration stress and to tolerate freezing stress may be related to the mechanical strength of the wall within *A. fistulosum*. Pectin is known to play a large role in the strength and stability of the cell wall [4].

### 4.3. Dehydration Stress in Arabidopsis and Pectin Proxy

Similar to calcium and HG, boron and RG-II have the ability to form cross-linkages, which results in the formation of RG-II dimers [4–6]. Previous research by Panter et al. (2019) examining the *MUR1* gene in *Arabidopsis* found that the formation of these RG-II dimers improves tolerance to freezing stress [24]. Since dehydration stress is associated with freezing stress, it is plausible that RG-II dimers positively influence the water holding capacity of pectin. Thus, the addition of boron to GENU BETA likely resulted in the formation of RG-II dimers since the 8% GENU BETA pectin with boron lost significantly ($p < 0.05$) less water after 6 h compared to 8% GENU BETA pectin without boron.

Boron was also found to influence dehydration stress resistance in *Arabidopsis* in the various boron-transporter mutants. While none of the observed changes in dehydration stress resistance were found to be significant ($p > 0.05$) amongst the three boron-transporter mutants explored, *bor1* lost the most water. Decreased resistance to dehydration stress in *bor1* compared to *nip5;1* and *nip6;1* may be the result of differences in the roles of the boron transporters and further speaks to the importance of understanding the specific nature of the transporters.

Briefly, NIP5;1 and NIP6;1 are localized in the plasma membrane and are involved in borate transport [36,37]. NIP5;1 is also involved in arsenite transport [38]. In comparison, BOR1 is localized in the cytoplasm, endosome and vacuole in addition to the plasma membrane [38]. BOR1 is also involved in a wider range of functions, including detection of nutrients, ion homeostasis and transmembrane transport, as well as borate transport and response to boron-containing substances [38]. Therefore, a mutation in BOR1 may be more deleterious. However, there is a lack of additional research exploring NIP, BOR or other boron transporters in relation to dehydration stress resistance. Nonetheless, a range of species, including maize, wheat and tomatoes, had enhanced drought stress tolerance as a result of boron application [22,23,82,83]. This further implicates the importance of boron, and likely its structural role in the ability to retain water.

In addition to calcium and boron, PMEIs are another important aspect of pectin modifications and cell wall structure. PMEIs provide negative feedback on PMEs, in turn controlling the ability for "egg-box" structure formation in addition to altering the cell wall in a range of other structural ways [7]. The family of PMEIs within *Arabidopsis* is broad, and currently includes 71 putative genes [7,84,85]. Despite a lack of significance ($p > 0.05$), our results showed an over-expression of PMEI5 was beneficial in resistance to dehydration stress, by way of reducing percentage water loss and electrolyte leakage.

While seemingly counterintuitive, given that the over-expression of PMEI would reduce the formation of "egg-box" structures, research from An et al. (2008) supports our findings [86]. They found the over-expression of CaPMEI1, a PMEI from peppers, enhanced tolerance to drought stress [86]. This was attributable to an increased tolerance to oxidative stress, which was observed in the *CaPMEI1-OX Arabidopsis* line, which may in turn reduce the damage caused by stresses such as drought [86]. However, findings by An et al. (2008) and those observed in this study stand in direct contrast to findings by from Amsbury et al. (2016) and Yang et al. (2019), who found increased de-methylation of pectin enhanced drought stress tolerance [86–88]. This may be the result of differences amongst the various PMEIs.

### 4.4. Fungal Pathogens

Aside from abiotic stresses, biotic stresses also pose an immense threat to plants. Since both categories of stress may occur at the same time, a common mechanism of defense is critical to survival. The cell wall itself is the first line of defense and often serves as a signaling mechanism alerting plants to disease through pattern recognition receptors (PRRs) and damage-associated molecular patterns (DAMPs) [89]. Pectin methylesterase (PME) has long been associated with facilitated viral movement through the association between viral-encoded mobility proteins (MPs) and PME in which PME specifically recognizes binding domains of the viral MPs [90]. Lionetti et al. (2014) hypothesized once bound, the PME-MP increase the diameter of the plasmodesmata through the loosening of the callose and thereby facilitating greater viral movement [91]. They proposed that the inhibitor, PMEI, inhibits viral movement due to the reduction in PME-MP, thereby limiting the plasmodesmata dilation [91]. While the PMEI5 in our experiments did not show a significant reduction in fungal spread, it could be due to differences in homologous traits between the various PMEs and PMEIs. In Arabidopsis, there are 71 known PMEI genes and 66 known PME genes [7,85,92]. Given the large number of PMEs, it is highly plausible that only specific PMEs interact with MPs. The ability for the cell wall to remodel in response to viral pathogens is another critical component in defense, as it increases the strength of the cell wall [93]. This process has been found to be an important mechanism of defense against a variety of viral pathogens including halo blight disease in common bean, and *Potato virus* [93–95]. PMEI and PME have both been found to mediate cell wall remodeling [96,97]. While this set of experiments did not specifically focus on cell wall modeling in relation to viruses, the role of PMEI5 in connection with cell wall remodeling and defense against *B. cinerea* and *C. higginsianum* should be explored further. The analysis of lesion size over the course of 5 dpi for both *B. cinerea* and *C. higginsianum* demonstrated boron plays an integral role in resistance against both pathogens. *bor1* was significantly ($p < 0.05$) less resistant to *B. cinerea* compared to the other genotypes of interest (*nip5;1*, *nip6;1*, *p35S::PMEI5* and Col-0). This is likely the result of structural changes within pectin, and more specifically RG-II, given the findings of Petrasch et al. (2019) [28]. *bor1* was also significantly ($p < 0.05$) less resistant to *C. higginsianum* compared to Col-0. At 5 dpi, 100% of the total leaf area from *bor1* leaves was covered in a *B. cinerea* lesion. In comparison, leaves obtained from the other genotypes had less than 72% of their total leaf area covered lesions.

Since the BOR1 transporter is more widely expressed within the plant compared to NIP5;1 and NIP6;1, the *bor1* genotype likely has a lower concentration of boron [36–38]. Thus, the reduced resistance to *B. cinerea* in *bor1* compared to *nip5;1* and *nip6;1* may be the result of a lower concentration of boron, as boron has been previously found to mitigate *B. cinerea* infections [31].

Similarly, at 5 dpi following inoculation with *C. higginsianum*, 63.1% of the total *bor1* leaf area were covered in *C. higginsianum* lesion, compared to an average of only 31.7% for the other genotypes of interest. In addition to having twice the total leaf area covered in a lesion at 5 dpi, as early as 1 dpi, the *bor1* total leaf area covered in a *C. higginsianum* lesion was over 30 times greater compared to the other genotypes of interest, suggesting boron transported through the BOR1 channel is vital in the resistance to this disease. It is well understood that boron plays a critical role in the structure of the cell wall and a mutation in BOR1 would likely reduce the concentration of boron within the cell wall [98].

The reduced function of NIP5;1 and NIP6;1, in addition to the likelihood of redundancy within the NIP family, may be responsible for differences in leaf tissue infections compared to BOR1.

Furthermore, the over-expression of PMEI1, PMEI2, PMEI10, PMEI11 and PMEI12 has been previously found to help to maintain the integrity of the cell wall in plants infected with *B. cinerea* [26,27]. However, the over-expression of PMEI5 in the *p35S::PMEI5* did not significantly ($p > 0.05$) alter resistance to *B. cinerea*. Instead, leaves obtained from the *p35S::PMEI5* had the second greatest leaf area covered in *B. cinerea* lesions, after *bor1*. The contrasting nature of these findings compared to those obtained during our experiments may be the result of variation amongst PMEIs. PMEI5 belongs to Group 3 of PMEIs while previously explored PMEIs belong to other groups [26]. Future studies should focus on gaining a greater understanding into the differences between Group 1, 2 and 3 PMEIs.

Moreover, the more substantial lesion spread at 1 dpi on *bor1* leaves suggests weaker cell walls within the *bor1* genotype. *C. higginsianum* penetrates the cell wall of the host through physical pressure without the use of enzymes, unlike *B. cinerea* [99]. The rapid growth of lesions on *bor1* leaves is indicative that *C. higginsianum* may have penetrated the cell walls with less pressure, signaling potential weakness of the *bor1* cell walls, likely the result of lower concentrations of boron and in turn, a reduction in the number of RG-II dimers. Differences in function for the BOR1 transporter in comparison to NIP5;1 and NIP6;1 may reflect the variation observed in the percentage leaf area infected. More generally, while boron has been previously reported to act as an antifungal agent for a range of species infected with *Colletotrichum graminicola*, to our knowledge, there are no studies exploring NIP, BOR or any other B-transporters in relation to *C. higginsianum* [33,34].

## 5. Conclusions

While modifications in pectin do not allow the cell wall to form an impassable barrier, these elements remain critical in the role of the cell wall as a barrier to stress. While our findings on cell wall rheology, shear force, calcium localization and analysis of water loss in pure pectin standards are indicative that "egg-box" structures and RG-II dimers likely formed, this alone was not sufficient in significantly ($p > 0.05$) improving dehydration stress tolerance. Variation was also observed with respect to resistance to the fungal pathogens of interest, suggesting that while boron and PMEI5 likely play a role in resistance to dehydration stress and fungal pathogens, the true picture appears to be far more nuanced. Overall, our findings demonstrate the cell wall is not a silver bullet to dehydration stress or fungal pathogens. Even so, the results of this study help to further our understanding of the role of the cell wall as a physical barrier to both abiotic and biotic stresses.

**Supplementary Materials:** The following supporting information can be downloaded at: https://www.mdpi.com/article/10.3390/plants11030385/s1. Figure S1: Generalized additive models showing the relationship between temperature (°C) and pectin viscosity (Pa.S), Figure S2: Analysis of green pixels from images obtained from *Allium fistulosum* epidermal cell layers stained with fluorescein diacetate, Figure S3: Analysis of green pixels from images obtained from *Allium cepa* epidermal cell layers stained with fluorescein diacetate; Table S1: Primers used to genotype *nip5;1-1*, *nip6;1-2* and *bor1-3*, Table S2: ANOVA for generalized additive models examining the relationship between temperature (°C) and pectin viscosity (Pa.S), Table S3: ANOVA analyzing the force required to shear *Allium fistulosum* sheaths, Table S4: ANOVA analyzing percent water loss in *Allium fistulosum* and *Allium cepa* sheaths, Table S5: ANOVA analyzing percent water loss in pectin solutions, Table S6: Tukey's test analyzing percent water loss in pectin solutions, Table S7: ANOVA examining percent viability based on electrolyte leakage following dehydration in *Allium fistulosum* and *Allium cepa*, Table S8: Two-sample *t*-tests analyzing the intensity of green pixels following staining with fluorescein diacetate in *Allium fistulosum* and *Allium cepa*, Table S9: ANOVA analyzing lesion size following *Botrytis cinerea* inoculation in *Arabidopsis*, Table S10: Tukey test analyzing lesion size following *Botrytis cinerea* inoculation in *Arabidopsis*, Table S11: ANOVA analyzing lesion size following *Colletotrichum higginsianum* inoculation in *Arabidopsis*, Table S12: Tukey test analyzing lesion size following *Colletotrichum higginsianum* inoculation in *Arabidopsis*.

**Author Contributions:** Conceptualization, K.K.T.; methodology, K.K.T., C.K., Y.Z.F., S.G. and Y.W.; software, A.D.F., M.L. and J.S.; formal analysis, A.D.F., M.L. and J.S.; investigation, A.D.F., Y.Z.F., S.W. and K.K.T., resources, K.K.T., Y.W., L.Q. and S.G.; data curation, A.D.F., M.L. and J.S.; writing—original draft preparation, A.D.F. and K.K.T.; writing—review and editing, A.D.F. and K.K.T.; visualization, K.K.T.; supervision, K.K.T., C.K., Y.Z.F., S.G. and Y.W.; project administration, K.K.T.; funding acquisition, K.K.T. All authors have read and agreed to the published version of the manuscript.

**Funding:** This work was supported by the Alexander Graham Bell Canadian Graduate Scholarship-Masters (CGS-M) from the National Science and Engineering Research Council (NSERC) of Canada awarded to A.D.F, in addition to an NSERC Discovery grant awarded to K.K.T (RGPIN-2018-05853).

**Institutional Review Board Statement:** Not applicable.

**Informed Consent Statement:** Not applicable.

**Data Availability Statement:** Data supporting the above findings are available within this paper and Supplementary Materials published online.

**Acknowledgments:** We thank Fatemeh Keivaninahr and Chi Diem Doan (University of Saskatchewan (USask), Canada) for their assistance with the rheometer and Phyllis Shand (USask, Canada) for her help with texture analysis, Na Liu (Canadian Light Source, Canada) for assistance with sample preparation for the synchrotron and Eric Lamb (USask, Canada) for guidance related to generalized additive models. We would also like to acknowledge Keristein Muller (Veristat, Canada) for the p35S::PMEI5 line and Hong Wang (USask, Canada) for assistance in ordering primers. Eldon Siemens (Agriculture Greenhouse, USask, Canada) in addition to other greenhouse staff were instrumental in the growth of Allium plants. We acknowledge fellow lab members Ian Willick and Eric Rae for support and assistance in the project. Finally, we acknowledge the University of Saskatchewan for their generosity, as this work was supported in part by various scholarships and bursaries. Part of the research described in this paper was performed at the Canadian Light Source, a national research facility of the University of Saskatchewan, which is supported by the Canada Foundation for Innovation (CFI), the Natural Sciences and Engineering Research Council (NSERC), the National Research Council (NRC), the Canadian Institutes of Health Research (CIHR), the Government of Saskatchewan, and the University of Saskatchewan. This research used resources of the Advanced Photon Source, an Office of Science User Facility operated for the U.S. Department of Energy (DOE) Office of Science by Argonne National Laboratory and was supported by the U.S. DOE under Contract No. DE-AC02-06CH11357, and the Canadian Light Source and its funding partners.

**Conflicts of Interest:** The authors declare no conflict of interest.

## Abbreviations

| | |
|---|---|
| AIC Score | Akaike Information Criterion (AIC). Used to select the optimal generalized additive model. A lower value indicates a more optimal model. |
| BOR1 | Boric acid channel: part of the BOR family. |
| CA | Calcium Applied/ Calcium Treatment. |
| Cell Wall | Within the context of this paper, this term also incorporates the middle lamella. |
| EDF | Empirical Distribution Factor. Distribution of a function, where a value of one is indicative of a linear relationship, while a value above one is indicative of a non-linear relationship. |
| FDA | Fluorescein Diacetate. |
| GAM | Generalized Additive Model. A generalized linear model where the effect of predictive variables is captured using smoothing functions. |
| GB | GENU BETA Pectin (Sugar Beet Pectin). |
| HG. | Homogalacturonan. |
| HM | High Methylated Citrus Pectin. |
| LM | Low Methylated Citrus Pectin. |
| NCA | No Calcium Applied/ No Calcium Treatment. |
| NIP5:1 | Boric acid channel, part of the NIP family. |
| NIP6:1 | Boric acid channel, part of the NIP family. |
| PMEI5 | Pectin methylesterase inhibitor five, part of the PMEI family. |
| RG-II | Rhamnogalacturonan II. |

## References

1. Keegstra, K. Plant Cell Walls. *Plant Physiol.* **2010**, *154*, 483–486. [CrossRef] [PubMed]
2. Houston, K.; Tucker, M.; Chowdhury, J.; Shirley, N.; Little, A. The Plant Cell Wall: A Complex and Dynamic Structure as Revealed by the Responses of Genes under Stress Conditions. *Front. Plant Sci.* **2016**, *7*, 984. [CrossRef] [PubMed]
3. Cosgrove, D.C.; Jarvis, M.C. Comparative structure and biomechanics of plant primary and secondary cell walls. *Front. Plant Sci.* **2012**, *3*, 204. [CrossRef]
4. Mohnen, D. Pectin structure and biosynthesis. *Curr. Opin. Plant Biol.* **2008**, *11*, 266–277. [CrossRef]
5. Braccini, I.; Pérez, S. Molecular Basis of $Ca^{2+}$-Induced Gelation in Alginates and Pectins: The Egg-Box Model Revisited. *Biomacromolecules* **2001**, *2*, 1089–1096. [CrossRef] [PubMed]
6. Ravanat, G.; Rinaudo, M. Investigation on oligo- and polygalacturonic acids by potentiometry and circular dichroism. *Biopolymers* **1980**, *19*, 2209–2222. [CrossRef]
7. Wormit, A.; Usadel, B. The Multifaceted Role of Pectin Methylesterase Inhibitors (PMEIs). *Int. J. Mol. Sci.* **2018**, *19*, 2878. [CrossRef]

8. Ishii, T.; Matsunaga, T.; Pellerin, P.; O'Neill, M.A.; Darvill, A.; Albersheim, P. The Plant Cell Wall Polysaccharide Rhamnogalacturonan II Self-assembles into a Covalently Cross-linked Dimer. *J. Biol. Chem.* **1999**, *274*, 13098–13104. [CrossRef]
9. Matoh, T. Boron in Plant Cell Walls. In *Plant and Soil*; Kluwer Academic Publishers: Berlin/Heidelberg, Germany, 1997; Volume 193, pp. 59–70. Available online: https://link.springer.com/content/pdf/10.1023%2FA%3A1004207824251.pdf (accessed on 10 April 2021). [CrossRef]
10. Fleischer, A.; O'Neill, M.A.; Ehwald, R. The Pore Size of Non-Graminaceous Plant Cell Walls Is Rapidly Decreased by Borate Ester Cross-Linking of the Pectic Polysaccharide Rhamnogalacturonan II. *Plant Physiol.* **1999**, *121*, 829–838. [CrossRef]
11. Ryden, P.; Sugimoto-Shirasu, K.; Smith, A.C.; Findlay, K.; Reiter, W.-D.; McCann, M.C. Tensile Properties of Arabidopsis Cell Walls Depend on Both a Xyloglucan Cross-Linked Microfibrillar Network and Rhamnogalacturonan II-Borate Complexes. *Plant Physiol.* **2003**, *132*, 1033–1040. [CrossRef] [PubMed]
12. Willats, W.; Orfila, C.; Limberg, G.; Buchholt, H.C.; van Alebeek, G.-J.W.M.; Voragen, A.G.J.; Marcus, S.E.; Christensen, T.M.I.E.; Mikkelsen, J.D.; Murray, B.S.; et al. Modulation of the Degree and Pattern of Methyl-esterification of Pectic Homogalacturonan in Plant Cell Walls. *J. Biol. Chem.* **2001**, *276*, 19404–19413. [CrossRef] [PubMed]
13. Thor, K. Calcium—Nutrient and Messenger. *Front. Plant Sci.* **2019**, *10*, 440. [CrossRef] [PubMed]
14. Alberts, B.; Johnson, A.; Lewis, J.; Raff, M.; Roberts, K.; Walter, P. The Plant Cell Wall. 2002. Available online: https://www.ncbi.nlm.nih.gov/books/NBK26928/ (accessed on 10 April 2021).
15. White, P.B.; Wang, T.; Park, Y.B.; Cosgrove, D.J.; Hong, M. Water–Polysaccharide Interactions in the Primary Cell Wall of Arabidopsis thaliana from Polarization Transfer Solid-State NMR. *J. Am. Chem. Soc.* **2014**, *136*, 10399–10409. [CrossRef]
16. Ha, M.-A.; Apperley, D.C.; Jarvis, M.C. Molecular Rigidity in Dry and Hydrated Onion Cell Walls. *Plant Physiol.* **1997**, *115*, 593–598. [CrossRef]
17. Mohnen, D.; Engle, K.; Amos, R.; Yan, J.Y.; Glushka, J.; Atmodjo, M.; Tan, L.; Moremen, K.W. A new model for pectin structural synthesis. In Proceedings of the Plant Cell Wall Biology 2021, Sapporo, Japan, 27 June–1 July 2021; p. 59.
18. Konno, H.; Yamasaki, Y.; Sugimoto, M.; Takeda, K. Differential changes in cell wall matrix polysaccharides and glycoside-hydrolyzing enzymes in developing wheat seedlings differing in drought tolerance. *J. Plant Physiol.* **2008**, *165*, 745–754. [CrossRef]
19. Leucci, M.R.; Lenucci, M.S.; Piro, G.; Dalessandro, G. Water stress and cell wall polysaccharides in the apical root zone of wheat cultivars varying in drought tolerance. *J. Plant Physiol.* **2008**, *165*, 1168–1180. [CrossRef]
20. Piro, G.; Leucci, M.R.; Waldron, K.; Dalessandro, G. Exposure to water stress causes changes in the biosynthesis of cell wall polysaccharides in roots of wheat cultivars varying in drought tolerance. *Plant Sci.* **2003**, *165*, 559–569. [CrossRef]
21. Herburger, K.; Xin, A.; Holzinger, A. Homogalacturonan Accumulation in Cell Walls of the Green Alga Zygnema sp. (Charophyta) Increases Desiccation Resistance. *Front. Plant Sci.* **2019**, *10*, 540. [CrossRef]
22. Abdel-Motagally, F.; El-Zohri, M. Improvement of wheat yield grown under drought stress by boron foliar application at different growth stages. *J. Saudi Soc. Agric. Sci.* **2018**, *17*, 178–185. [CrossRef]
23. Karim, M.R.; Zhang, Y.Q.; Zhao, R.R.; Chen, X.P.; Zhang, F.S.; Zou, C.Q. Alleviation of drought stress in winter wheat by late foliar application of zinc, boron, and manganese. *J. Plant Nutr. Soil Sci.* **2012**, *175*, 142–151. [CrossRef]
24. Panter, P.E.; Kent, O.; Dale, M.; Smith, S.J.; Skipsey, M.; Thorlby, G.; Cummins, I.; Ramsay, N.; Begum, R.A.; Sanhueza, D.; et al. MUR1-mediated cell-wall fucosylation is required for freezing tolerance in Arabidopsis thaliana. *New Phytol.* **2019**, *224*, 1518–1531. [CrossRef] [PubMed]
25. Lionetti, V.; Raiola, A.; Camardella, L.; Giovane, A.; Obel, N.; Pauly, M.; Favaron, F.; Cervone, F.; Bellincampi, D. Overexpression of Pectin Methylesterase Inhibitors in Arabidopsis Restricts Fungal Infection by Botrytis cinerea. *Plant Physiol.* **2007**, *143*, 1871–1880. [CrossRef] [PubMed]
26. Lionetti, V.; Fabri, E.; De Caroli, M.; Hansen, A.R.; Willats, W.G.; Piro, G.; Bellincampi, D. Three Pectin Methylesterase Inhibitors Protect Cell Wall Integrity for Arabidopsis Immunity to Botrytis. *Plant Physiol.* **2017**, *173*, 1844–1863. [CrossRef]
27. Raiola, A.; Lionetti, V.; Elmaghraby, I.; Immerzeel, P.; Mellerowicz, E.J.; Salvi, G.; Cervone, F.; Bellincampi, D. Pectin Methylesterase Is Induced in Arabidopsis upon Infection and Is Necessary for a Successful Colonization by Necrotrophic Pathogens. *Mol. Plant-Microbe Interact.* **2011**, *24*, 432–440. [CrossRef] [PubMed]
28. Petrasch, S.; Silva, C.J.; Mesquida-Pesci, S.D.; Gallegos, K.; van den Abeele, C.; Papin, V.; Fernandez-Acero, F.J.; Knapp, S.J.; Blanco-Ulate, B. Infection Strategies Deployed by Botrytis cinerea, Fusarium acuminatum, and Rhizopus stolonifer as a Function of Tomato Fruit Ripening Stage. *Front. Plant Sci.* **2019**, *10*, 223. [CrossRef]
29. Engelsdorf, T.; Will, C.; Hofmann, J.; Schmitt, C.; Merritt, B.B.; Rieger, L.; Frenger, M.S.; Marschall, A.; Franke, R.B.; Pattathil, S.; et al. Cell wall composition and penetration resistance against the fungal pathogenColletotrichum higginsianumare affected by impaired starch turnover in Arabidopsis mutants. *J. Exp. Bot.* **2016**, *68*, 701–713. [CrossRef]
30. Qin, G.; Zong, Y.; Chen, Q.; Hua, D.; Tian, S. Inhibitory effect of boron against Botrytis cinerea on table grapes and its possible mechanisms of action. *Int. J. Food Microbiol.* **2010**, *138*, 145–150. [CrossRef]
31. Wójcik, P.; Lewandowski, M. Effect of Calcium and Boron Sprays on Yield and Quality of "Elsanta" Strawberry. *J. Plant Nutr.* **2003**, *26*, 671–682. [CrossRef]
32. Shi, X.-Q.; Li, B.-Q.; Qin, G.-Z.; Tian, S.-P. Antifungal Activity and Possible Mode of Action of Borate Against Colletotrichum gloeosporioides on Mango. *Plant Dis.* **2011**, *95*, 63–69. [CrossRef]
33. Shi, X.; Li, B.; Qin, G.; Tian, S. Mechanism of antifungal action of borate against Colletotrichum gloeosporioides related to mitochondrial degradation in spores. *Postharvest Biol. Technol.* **2012**, *67*, 138–143. [CrossRef]

34. Palta, J.P.; Levitt, J.; Stadelmann, E.J. Freezing Injury in Onion Bulb Cells. *Plant Physiol.* **1977**, *60*, 393–397. [CrossRef] [PubMed]
35. Tanino, K.K.; Kobayashi, S.; Hyett, C.; Hamilton, K.; Liu, J.; Li, B.; Borondics, F.; Pedersen, T.; Tse, J.; Ellis, T.; et al. Allium fistulosumas a novel system to investigate mechanisms of freezing resistance. *Physiol. Plant.* **2012**, *147*, 101–111. [CrossRef] [PubMed]
36. TAIR. Locus: AT1G80760. 2013. Available online: https://www.arabidopsis.org/servlets/TairObject?name=at1g80760&type=locus (accessed on 10 April 2021).
37. TAIR. Locus: AT4G10380. 2015. Available online: https://www.arabidopsis.org/servlets/TairObject?type=locus&name=At4g10380 (accessed on 10 April 2021).
38. TAIR. Locus: AT2G47160. 2021. Available online: https://www.arabidopsis.org/servlets/TairObject?name=AT2G47160&type=locus (accessed on 10 April 2021).
39. Müller, K.; Levesque-Tremblay, G.; Bartels, S.; Weitbrecht, K.; Wormit, A.; Usadel, B.; Haughn, G.; Kermode, A.R. Demethylesterification of Cell Wall Pectins in Arabidopsis Plays a Role in Seed Germination. *Plant Physiol.* **2012**, *161*, 305–316. [CrossRef] [PubMed]
40. Edwards, K.; Johnstone, C.; Thompson, C. A simple and rapid method for the preparation of plant genomic DNA for PCR analysis. *Nucleic Acids Res.* **1991**, *19*, 1349. [CrossRef] [PubMed]
41. Wang, S.; Wen, R.; Shi, X.; Lambrecht, A.; Wang, H.; Xiao, W. RAD5a and REV3 function in two alternative pathways of DNA-damage tolerance in Arabidopsis. *DNA Repair* **2011**, *10*, 620–628. [CrossRef]
42. Hutton, B.; (TIC Gums, White Marsh, MD, USA). Personal Communication, 2019.
43. Meyers, D.; (CP Kelco, Atlanta, GA, USA). Personal Communication, 2019.
44. Yapo, B.M.; Lerouge, P.; Thibault, J.-F.; Ralet, M.-C. Pectins from citrus peel cell walls contain homogalacturonans homogenous with respect to molar mass, rhamnogalacturonan I and rhamnogalacturonan II. *Carbohydr. Polym.* **2007**, *69*, 426–435. [CrossRef]
45. Mankarios, A.T.; Hall, M.A.; Jarvis, M.C.; Threlfall, D.R.; Friend, J. Cell wall polysaccharides from onions. *Phytochemistry* **1980**, *19*, 1731–1733. [CrossRef]
46. Willick, I.R. The Mechanism of Freezing Resistance in Cold-Acclimated Winter Wheat and Rye Crowns. Ph.D. Thesis, The University of Saskatchewan, Saskatoon, SK, Canada, 2018.
47. Flint, H.L.; Boyce, B.R.; Beattie, D.J. Index of injury—A useful expression of freezing injury to plant tissues as determined by the electrolytic method. *Can. J. Plant Sci.* **1967**, *47*, 229–230. [CrossRef]
48. Banik, P.; Zeng, W.; Tai, H.; Bizimungu, B.; Tanino, K. Effects of drought acclimation on drought stress resistance in potato (Solanum tuberosum L.) genotypes. *Environ. Exp. Bot.* **2016**, *126*, 76–89. [CrossRef]
49. Stobbs, J.; (Canadian Light Source, Saskatoon, SK, Canada). Personal Communication, 2021.
50. Sole, V.A.; Papillon, E.; Cotte, M.; Walter, P.; Susini, J. A multiplatform code for the analysis of energy-dispersive X-ray fluorescence spectra. *Spectrochim. Acta Part B At. Spectrosc.* **2007**, *62*, 63–68. [CrossRef]
51. Liu, G.; Ji, Y.; Bhuiyan, N.H.; Pilot, G.; Selvaraj, G.; Zou, J.; Wei, Y. Amino Acid Homeostasis Modulates Salicylic Acid–Associated Redox Status and Defense Responses inArabidopsis. *Plant Cell* **2010**, *22*, 3845–3863. [CrossRef] [PubMed]
52. Liu, G.; Kennedy, R.; Greenshields, D.L.; Peng, G.; Forseille, L.; Selvaraj, G.; Wei, Y. Detached and Attached *Arabidopsis* Leaf Assays Reveal Distinctive Defense Responses Against Hemibiotrophic *Colletotrichum* spp. *Mol. Plant-Microbe Interact.* **2007**, *20*, 1308–1319. [CrossRef] [PubMed]
53. Wang, S.; Li, Q.; Zhao, L.; Fu, S.; Qin, L.; Wei, Y.; Fu, Y.-B.; Wang, H. Arabidopsis UBC22, an E2 able to catalyze lysine-11 specific ubiquitin linkage formation, has multiple functions in plant growth and immunity. *Plant Sci.* **2020**, *297*, 110520. [CrossRef] [PubMed]
54. Wood, S.N. *Generalized Additive Models: An Introduction with R*, 2nd ed.; CRC Press: Boca Raton, FL, USA, 2017.
55. Wood, S. *Package 'mgcv'*; 1.8-38. CRAN, 2021. Available online: https://cran.r-project.org/web/packages/mgcv/mgcv.pdf (accessed on 20 April 2021).
56. Adler, D.; Murdoch, D. rgl: 3D Visualization Using OpenGL. CRAN. 4 March 2021. Available online: https://rdrr.io/cran/rgl/ (accessed on 20 April 2021).
57. Fasiolo, M.; Nedellec, R.; Goude, Y.; Capezza, C.; Wood, S. *mgcViz: Visualisations for Generalized Additive Models*; 0.1.9. CRAN, 2020. Available online: https://cran.r-project.org/web/packages/mgcViz/mgcViz.pdf (accessed on 20 April 2021).
58. Wickham, H.; Hester, J.; Chang, W. *Devtools: Tools to Make Developing R Packages Easier*; 2.4.3. CRAN, 2021. Available online: https://cran.r-project.org/web/packages/devtools/devtools.pdf (accessed on 20 April 2021).
59. Polymer Solutions News Team. Metal Properties: Hardness, Toughness, & Strength. 2015. Available online: https://www.polymersolutions.com/blog/defining-metal-properties/ (accessed on 20 April 2021).
60. Ritchie, R.O. The conflicts between strength and toughness. *Nat. Mater.* **2011**, *10*, 817–822. [CrossRef] [PubMed]
61. O'Neill, M.A.; Ishii, T.; Albersheim, P.; Darvill, A.G. Rhamnogalacturonan ii: Structure and function of a borate cross-linked cell wall pectic polysaccharide. *Annu. Rev. Plant Biol.* **2004**, *55*, 109–139. [CrossRef]
62. Hu, W.; Tian, S.B.; Di, Q.; Duan, S.H.; Dai, K. Effects of exogenous calcium on mesophyll cell ultrastructure, gas exchange, and photosystem II in tobacco (*Nicotiana tabacum* Linn.) under drought stress. *Photosynthetica* **2018**, *56*, 1204–1211. [CrossRef]
63. Hosseini, S.A.; Réthoré, E.; Pluchon, S.; Ali, N.; Billiot, B.; Yvin, J.-C. Calcium Application Enhances Drought Stress Tolerance in Sugar Beet and Promotes Plant Biomass and Beetroot Sucrose Concentration. *Int. J. Mol. Sci.* **2019**, *20*, 3777. [CrossRef]

64. Khushboo; Bhardwaj, K.; Singh, P.; Raina, M.; Sharma, V.; Kumar, D. Exogenous application of calcium chloride in wheat genotypes alleviates negative effect of drought stress by modulating antioxidant machinery and enhanced osmolyte accumulation. *Vitr. Cell. Dev. Biol. Anim.* **2018**, *54*, 495–507. [CrossRef]
65. Schatzberg, P. Molecular diameter of water from solubility and diffusion measurements. *J. Phys. Chem.* **1967**, *71*, 4569–4570. [CrossRef]
66. Liu, J. *Temperature-Mediated Alterations of the Plant Apoplast as a Mechanism of Intracellular Freezing Stress Avoidance*; University of Saskatchewan: Saskatoon, SK, Canada, 2015; Available online: https://harvest.usask.ca/handle/10388/ETD-2015-10-2290 (accessed on 20 April 2021).
67. Trewavas, A.; Knight, M. Mechanical signalling, calcium and plant form. In *Signals and Signal Transduction Pathways in Plants*; Springer: Dordrecht, The Netherlands, 1994; pp. 93–105. [CrossRef]
68. Knight, H.; Trewavas, A.J.; Knight, M. Calcium signalling in Arabidopsis thaliana responding to drought and salinity. *Plant J.* **1997**, *12*, 1067–1078. [CrossRef] [PubMed]
69. Wang, J.-P.; Munyampundu, J.-P.; Xu, Y.-P.; Cai, X.-Z. Phylogeny of Plant Calcium and Calmodulin-Dependent Protein Kinases (CCaMKs) and Functional Analyses of Tomato CCaMK in Disease Resistance. *Front. Plant Sci.* **2015**, *6*, 1075. [CrossRef] [PubMed]
70. Bender, K.W.; Snedden, W.A. Calmodulin-Related Proteins Step Out from the Shadow of Their Namesake. *Plant Physiol.* **2013**, *163*, 486–495. [CrossRef] [PubMed]
71. Luan, S. The CBL–CIPK network in plant calcium signaling. *Trends Plant Sci.* **2009**, *14*, 37–42. [CrossRef] [PubMed]
72. Yang, T.; Poovaiah, B. Calcium/calmodulin-mediated signal network in plants. *Trends Plant Sci.* **2003**, *8*, 505–512. [CrossRef]
73. Cheng, S.-H.; Willmann, M.R.; Chen, H.-C.; Sheen, J. Calcium Signaling through Protein Kinases. The Arabidopsis Calcium-Dependent Protein Kinase Gene Family. *Plant Physiol.* **2002**, *129*, 469–485. [CrossRef]
74. Hepler, P.K. Calcium: A Central Regulator of Plant Growth and Development. *Plant Cell* **2005**, *17*, 2142–2155. [CrossRef]
75. Burstrom, H.G. Calcium and plant growth. *Biol. Rev.* **1968**, *43*, 287–316. [CrossRef]
76. Jones, R.G.W.; Lunt, O.R. The function of calcium in plants. *Bot. Rev.* **1967**, *33*, 407–426. [CrossRef]
77. Arora, R. Mechanism of freeze-thaw injury and recovery: A cool retrospective and warming up to new ideas. *Plant Sci.* **2018**, *270*, 301–313. [CrossRef]
78. Moore, J.P.; Vicré-Gibouin, M.; Farrant, J.M.; Driouich, A. Adaptations of higher plant cell walls to water loss: Drought vs desiccation. *Physiol. Plant.* **2008**, *134*, 237–245. [CrossRef] [PubMed]
79. Burke, M.J.; Gusta, L.V.; Quamme, A.H.; Weiser, C.J.; Li, P.H. Freezing and Injury in Plants. *Annu. Rev. Plant Physiol.* **1976**, *27*, 507–528. [CrossRef]
80. Oertli, J.J. Negative Turgor Pressures in Plant Cells. *Z. Für Pflanz. Und Bodenkd.* **1986**, *149*, 60–67. [CrossRef]
81. Oertli, J.J.; Lips, S.H.; Agami, M. The strength of sclerophyllous cells to resist collapse due to negative turgor pressure. *Acta Oecol.* **1990**, *11*, 281–289.
82. Aydin, M.; Tombuloglu, G.; Sakcali, M.S.; Hakeem, K.R.; Tombuloglu, H. Boron Alleviates Drought Stress by Enhancing Gene Expression and Antioxidant Enzyme Activity. *J. Soil Sci. Plant Nutr.* **2019**, *19*, 545–555. [CrossRef]
83. Naeem, M.; Ahmad, R.; Ahmad, R.; Ashraf, M.Y.; Ihsan, M.Z.; Nawaz, F.; Athar, H.-U.; Abbas, H.T.; Abdullah, M. Improving drought tolerance in maize by foliar application of boron: Water status, antioxidative defense and photosynthetic capacity. *Arch. Agron. Soil Sci.* **2017**, *64*, 626–639. [CrossRef]
84. Xue, C.; Guan, S.-C.; Chen, J.-Q.; Wen, C.-J.; Cai, J.-F.; Chen, X. Genome wide identification and functional characterization of strawberry pectin methylesterases related to fruit softening. *BMC Plant Biol.* **2020**, *20*, 13–17. [CrossRef]
85. Wang, M.; Yuan, D.; Gao, W.; Li, Y.; Tan, J.; Zhang, X. A Comparative Genome Analysis of PME and PMEI Families Reveals the Evolution of Pectin Metabolism in Plant Cell Walls. *PLoS ONE* **2013**, *8*, e72082. [CrossRef]
86. An, S.H.; Sohn, K.H.; Choi, H.W.; Hwang, I.S.; Lee, S.C.; Hwang, B.K. Pepper pectin methylesterase inhibitor protein CaPMEI1 is required for antifungal activity, basal disease resistance and abiotic stress tolerance. *Planta* **2008**, *228*, 61–78. [CrossRef]
87. Amsbury, S.; Hunt, L.; Elhaddad, N.; Baillie, A.; Lundgren, M.; Verhertbruggen, Y.; Scheller, H.; Knox, P.; Fleming, A.J.; Gray, J.E. Stomatal Function Requires Pectin De-methyl-esterification of the Guard Cell Wall. *Curr. Biol.* **2016**, *26*, 2899–2906. [CrossRef]
88. Yang, W.; Ruan, M.; Xiang, M.; Deng, A.; Du, J.; Xiao, C. Overexpression of a pectin methylesterase gene PtoPME35 from Populus tomentosa influences stomatal function and drought tolerance in Arabidopsis thaliana. *Biochem. Biophys. Res. Commun.* **2019**, *523*, 416–422. [CrossRef] [PubMed]
89. Bacete, L.; Mélida, H.; Miedes, E.; Molina, A. Plant cell wall-mediated immunity: Cell wall changes trigger disease resistance responses. *Plant J.* **2018**, *93*, 614–636. [CrossRef] [PubMed]
90. Chen, M.-H.; Sheng, J.; Hind, G.; Handa, A.K.; Citovsky, V. Interaction between the tobacco mosaic virus movement protein and host cell pectin methylesterases is required for viral cell-to-cell movement. *EMBO J.* **2000**, *19*, 913–920. [CrossRef] [PubMed]
91. Lionetti, V.; Raiola, A.; Cervone, F.; Bellincampi, D. How do pectin methylesterases and their inhibitors affect the spreading of tobamovirus? *Plant Signal. Behav.* **2014**, *9*, e972863. [CrossRef]
92. Bethke, G.; Grundman, R.E.; Sreekanta, S.; Truman, W.; Katagiri, F.; Glazebrook, J. Arabidopsis PECTIN METHYLESTERASEs Contribute to Immunity against Pseudomonas syringae. *Plant Physiol.* **2013**, *164*, 1093–1107. [CrossRef]
93. De la Rubia, A.; Mélida, H.; Centeno, M.; Encina, A.; García-Angulo, P. Immune Priming Triggers Cell Wall Remodeling and Increased Resistance to Halo Blight Disease in Common Bean. *Plants* **2021**, *10*, 1514. [CrossRef]

94. Otulak-Kozieł, K.; Kozieł, E.; Lockhart, B.E.L.; Bujarski, J.J. The Expression of Potato Expansin A3 (StEXPA3) and Extensin4 (StEXT4) Genes with Distribution of StEXPAs and HRGPs-Extensin Changes as an Effect of Cell Wall Rebuilding in Two Types of PVYNTN–Solanum tuberosum Interactions. *Viruses* **2020**, *12*, 66. [CrossRef]
95. Otulak-Kozieł, K.; Kozieł, E.; Lockhart, B.E.L. Plant Cell Wall Dynamics in Compatible and Incompatible Potato Response to Infection Caused by Potato Virus Y (PVYNTN). *Int. J. Mol. Sci.* **2018**, *19*, 862. [CrossRef]
96. Willats, W.G.; McCartney, L.; Mackie, W.; Knox, J.P. Pectin: Cell biology and prospects for functional analysis. *Plant Mol. Biol.* **2001**, *47*, 9–27. [CrossRef]
97. Baluška, F.; Liners, F.; Hlavačka, A.; Schlicht, M.; Van Cutsem, P.; McCurdy, D.W.; Menzel, D. Cell wall pectins and xyloglucans are internalized into dividing root cells and accumulate within cell plates during cytokinesis. *Protoplasma* **2005**, *225*, 141–155. [CrossRef]
98. Brown, P.H.; Bellaloui, N.; Wimmer, M.A.; Bassil, E.S.; Ruiz, J.; Hu, H.; Pfeffer, H.; Dannel, F.; Römheld, V. Boron in Plant Biology. *Plant Biol.* **2002**, *4*, 205–223. [CrossRef]
99. Yan, Y.; Yuan, Q.; Tang, J.; Huang, J.; Hsiang, T.; Wei, Y.; Zheng, L. Colletotrichum higginsianum as a Model for Understanding Host–Pathogen Interactions: A Review. *Int. J. Mol. Sci.* **2018**, *19*, 2142. [CrossRef] [PubMed]

*plants*

MDPI

*Article*

# Immune Priming Triggers Cell Wall Remodeling and Increased Resistance to Halo Blight Disease in Common Bean

**Alfonso Gonzalo De la Rubia, Hugo Mélida , María Luz Centeno, Antonio Encina and Penélope García-Angulo ***

Área de Fisiología Vegetal, Departamento de Ingeniería y Ciencias Agrarias, Universidad de León, E-24071 León, Spain; algor@unileon.es (A.G.D.l.R.); hmelm@unileon.es (H.M.); mlcenm@unileon.es (M.L.C.); a.encina@unileon.es (A.E.)

* Correspondence: pgara@unileon.es

**Abstract:** The cell wall (CW) is a dynamic structure extensively remodeled during plant growth and under stress conditions, however little is known about its roles during the immune system priming, especially in crops. In order to shed light on such a process, we used the *Phaseolus vulgaris-Pseudomonas syringae* (Pph) pathosystem and the immune priming capacity of 2,6-dichloroisonicotinic acid (INA). In the first instance we confirmed that INA-pretreated plants were more resistant to Pph, which was in line with the enhanced production of $H_2O_2$ of the primed plants after elicitation with the peptide flg22. Thereafter, CWs from plants subjected to the different treatments (non- or Pph-inoculated on non- or INA-pretreated plants) were isolated to study their composition and properties. As a result, the Pph inoculation modified the bean CW to some extent, mostly the pectic component, but the CW was as vulnerable to enzymatic hydrolysis as in the case of non-inoculated plants. By contrast, the INA priming triggered a pronounced CW remodeling, both on the cellulosic and non-cellulosic polysaccharides, and CW proteins, which resulted in a CW that was more resistant to enzymatic hydrolysis. In conclusion, the increased bean resistance against Pph produced by INA priming can be explained, at least partially, by a drastic CW remodeling.

**Keywords:** bean; *Pseudomonas;* cell wall; plant defense; disease resistance; 2,6-dichloroisonicotinic acid; INA

**Citation:** De la Rubia, A.G.; Mélida, H.; Centeno, M.L.; Encina, A.; García-Angulo, P. Immune Priming Triggers Cell Wall Remodeling and Increased Resistance to Halo Blight Disease in Common Bean. *Plants* **2021**, *10*, 1514. https://doi.org/10.3390/plants10081514

Academic Editor: William Underwood

Received: 30 June 2021
Accepted: 19 July 2021
Published: 23 July 2021

**Publisher's Note:** MDPI stays neutral with regard to jurisdictional claims in published maps and institutional affiliations.

**Copyright:** © 2021 by the authors. Licensee MDPI, Basel, Switzerland. This article is an open access article distributed under the terms and conditions of the Creative Commons Attribution (CC BY) license (https://creativecommons.org/licenses/by/4.0/).

## 1. Introduction

In nature, plants are usually under biotic stress caused by different pathogens, making their survival difficult [1]. To combat attackers, plants have developed a sophisticated immune system, whose activation usually produces a decrease in yield and fitness [2], and therefore it must be tightly regulated. In this immune system, the plant cell wall (CW) is the first physical barrier to prevent the pathogen from progressing into the cell, having the capacity for remodeling the architecture of its components and their physicochemical properties [3–5].

Plant CW is a dynamic structure which surrounds every plant cell, determining their shape and providing mechanical support to the protoplasts to counteract the turgor pressure. Elongating cells are walled by the primary CW, a thin layer in continuous change, mainly composed of a complex matrix of cellulose, pectins, and hemicelluloses, and structural glycoproteins [6,7]. Cellulose is the main CW scaffold, which is composed of numerous β-(1-4)-glucans tightly packed to form microfibrils, very resistant to enzymatic hydrolysis. The hemicelluloses and pectins, commonly named as matrix polysaccharides, are a group of several distinct polysaccharides [7]. Pectins are a group of complex acid polysaccharides that are divided in three main domains by their occurrence: homogalacturonan (HG), rhamnogalacturonan I (RG-I), and rhamnogalacturonan II (RG-II). The proportion of these pectin domains depends on species, tissue, and even cell type, with the presence of minor pectic domains such as apiogalacturonan or xylogalacturonan. The binding among these pectins results in complex macromolecular structures which form a hydrated gel phase where cellulose and hemicelluloses are embedded [8]. They are mostly extracted from the CW by means of calcium chelators or hot water treatments. Hemicelluloses usually bind to the surface of the cellulose microfibrils, often bearing short side branches, and are typically extracted under alkali incubation. Dominant hemicelluloses in primary CW are xyloglucan (XG), xylan, or arabinoxylan (AX) depending on plant species [9]. Accompanying polysaccharides, structural proteins such as hydroxyproline-rich glycoproteins (HRGP) or arabinogalactan proteins (AGPs) are widespread in the CW matrix [10,11]. When the cell stops growing, a strengthened secondary CW, mainly composed of cellulose, hemicelluloses, and lignin, is deposited between preexisting primary CW and the plasma membrane [12,13].

As plant morphogenesis ultimately depends on CW properties, this structure needs to adapt to any cellular change at any developmental stage along the entire life of a plant [10]. These CW processes are tightly regulated to control the metabolism responsible of CW remodeling. Firstly, the synthesis loci differ depending on the CW polymers. Cellulose is synthesized at the plasma membrane [14], while matrix polysaccharides are deposited throughout the secretory apparatus [15,16]. Secondly, the properties that these polymers initially could confer to the matrix can be changed once they are deposited, by homo and/or hetero-crosslinking and by modifying the pattern and/or degree of backbone substitution. For instance, HG is synthesized in Golgi apparatus with most of its acidic carboxyl groups methyl-esterified, however later on, by removing the methyl groups, ionic bridges can be formed by calcium ions, establishing a network among HG chains [8]. RG-I and RG-II, which are pectins decorated with neutral sugars chains, are complex polysaccharides whose abundance, cross-linking and substitutions are also regulated in muro, depending on the development stage [8,17]. To gain resistance, RG-I backbone can be enlarged, rendering a higher molecular weight of the polysaccharides. This can be also achieved by cross-linking throughout borate diesters between the RG-II domains [17]. On the other hand, typical modifications of dicotyledonous hemicellulosic glycans are the presence of fucose or galactose residues attached to the XG backbone, which is important during the cell elongation [18]. In the same way, extensins, which are HRGP, play a crucial role in the structural and compositional CW plasticity, by forming intra- and intermolecular cross-links [19].

CW remodeling serves to strengthen the CW, maintaining the plasticity to react to environmental changes, being especially important under a pathogen attack. The role of the CW components in defense has been attested in different works carried out with several mutants mainly of the model species *Arabidopsis thaliana* (Arabidopsis), in which a specific CW alteration translated into resistance or vulnerability against different diseases [20]. For example, the impairment of cellulose synthesis in an isoxaben-resistant mutant (*ixr1*, also known as *cev1*) resulted in an enhanced resistance to fungal pathogens by means of constitutive activation of some defensive pathways [21]. Similarly, Arabidopsis mutants unable to synthesize cellulose in the secondary CW, as *irx1-6*, showed resistance against the fungus *Plectosphaerella cucumerina* and the bacterium *Ralstonia pseudosolanacearum* [22]. The impairment in the callose deposition, observed in constitutively activated cell-death (*cad1-3*) Arabidopsis mutants, produced an increment to *Pseudomonas syringae* susceptibility [23]. In the case of lignin, the overexpression of *WRKY1* gene in rice, which increased lignification and resulted in major resistance against *Verticillium dahlia* [24]. By contrast, downregulation of the genes involved in lignin biosynthesis (hydroxycinnamoyl-CoA shikimate quinate hydroxycinnamoyl transferase -HCT- or cinnamoyl-CoA reductase 1 -CCR1-), with alteration in lignin content and composition, triggered defense responses in Arabidopsis [25]. Interestingly, the role of polysaccharide feruloylation in monocot *Brachypodium* and dicot Arabidopsis was also demonstrated by the transgenic expression of a fungal feruloyl esterase in both species, which reduced mono and/or dimeric ferulic acids, resulting in major susceptibility to *Bipolaris sorokiniana* and *Botrytis cinerea*, respectively [26].

Similar effects were found when matrix polysaccharides were altered. Arabidopsis mutants repressed in glucuronate 4-epimerases genes (*gae1* and *gae6*) had an impairment in galacturonic acid and showed less resistance to *P. syringae* pv. *maculicola* and *Botrytis cinerea* [27]. In addition, Arabidopsis response regulator 6 mutant (*arr6*) was enriched in pectins and showed more resistance to *P. cucumerina* but more susceptibility to *R. pseudosolanacearum* [28]. Reductions in the content of galactomannans and in the decoration of xyloglucan, observed in immutans (*im*) variegation of Arabidopsis, involved susceptibility to *P. syringae* [29]. Finally, with respect to CW proteins, the reduced residual arabinose-2 (*rra2*) mutant, defective in extensin arabinosylation, could limit oomycete colonization in Arabidopsis roots [19]. Other CW enzymes known to participate in pathogenesis are pectin modified enzymes, such as pectin methylesterases (PME), which take part in the demethylesterification of HG. An example is the infection of *P. syringae pv. macuolicola* in Arabidopsis, which leads to an increment of PME activity [27]. Thus, *pme17* Arabidopsis mutants showed an increase in susceptibility to *B. cinerea* [30]. In the same way, PME inhibitor (PMEI) mutants such as *pmei10*, *pmei11*, and *pmei12* lead to immunity against *B. cinerea* [31].

Apart from mutants, another strategy consists in comparing resistant and susceptible wheat breeding lines to *Fusarium graminearum*. It was revealed that resistance could be due to CW changes such as a higher content of lignin, a higher degree of arabinosylation on xylans, and a higher degree and different pattern of methylesterification on pectins [32].

All these works were based on CW modified mutants or breeding lines that showed susceptibility or resistance against pathogens, but their CW was already modified on these genotypes, so the level of susceptibility was determined by a preformed structure that in some mutants involved defensive pathways constitutively activated [20]. However, much less is known about CW remodeling at a physiological stage after pathogen infection. In this sense, most studies have focused on interactions with necrotrophic fungi or insects, which produces mechanical damage in the CW by the release of CW degrading enzymes (CWDEs) [26,32]. The main documented responses so far are the local deposition of CW materials, known as papillae or the cross-linking among components of the matrix, in order to hinder the access of the pathogens to the protoplast [22,31,32]. Recently, several transcriptomic and proteomic analysis would indicate that CW remodeling is a widely extended process in nature [33–36], however the detailed characterization of the changes that take place in muro after bacteria attack remain poorly characterized.

Among other defense layers, it has been proposed that cellular defense responses can be activated, but stronger, faster, and more efficiently, by means of the application of different treatments, a phenomenon known as immune priming [37]. However, it is not known whether, somehow, this immune priming conducts to the strengthening of the CW, or the immunity triggers a remodeled CW capable to cope with a second attack from pathogens. Over the past decades, several stimuli, compounds, or abiotic stresses have been studied in order to produce priming [37–41]. This new state is reached by changes at physiological, transcriptional, metabolic, and epigenetic levels, which can hold priming throughout the entire life of a plant and can even be inherited by further generations [42,43]. Among other compounds proposed to produce this long-term effect, it is the functional salycilic acid (SA) analogue 2,6-dichloroisonicotinic acid (INA), the first synthetic compound to induce defense responses in laboratory [39]. Recent studies have proposed that INA enhances basal disease resistance in common bean (*Phaseolus vulgaris* L.) by epigenetic changes that can even be inherited [41,43]. However, although it has been described that the priming activator (R)-β-homoserine increased Arabidopsis CW defense against *H. arabidopsidis* by enhanced callose deposition [41], there are no studies that focus on CW remodeling after INA priming.

In this study, *Riñón* common bean was used. It is a widely cultivated variety in León (Spain), under the protected geographic identification (PGI) "Alubia de la Bañeza-León". This variety is attacked by the biotrophic gamma-proteobacteria *Pseudomonas syringae* pv. *phaseolicola* (Pph), causing halo blight disease [44–46], which causes yield losses of up to 45%, with its main symptoms being general chlorosis, stunting and distortion of growth [45]. Until now, disease control has mainly consisted in growing healthy seeds every season or replacing the susceptible varieties with others more resistant. This latter alternative is not suitable when the susceptible variety cultivated has gastronomical and economic interest, such as *Riñón* [45]. *Riñón* common bean has been previously described as a Pph-susceptible variety [44], although the reasons behind this susceptibility are unknown. A recent work from our group revealed that this bean variety is able to perceive the presence of Pph but it is unable to produce an effective and fast defense response, at least in part because of the lack of a quick SA peak production [47]. This is the reason why the use of SA analogue, such INA, could help in the defense process against Pph. On the other hand, there is no knowledge about CW remodeling after the Pph infection and it becomes relevant because Pph is a CWDEs producer [48].

Taking all of this into account, the aims of the present work were to know: (1), if a susceptible bean variety can remodel its CW after Pph inoculation, (2) if the application of a priming compound, such as INA, has effects in CW remodeling, and finally, (3) if these CW changes increased bean resistance against Pph. For these purposes, 15-day-old bean plants were pretreated or not with INA, and 1 week after they were inoculated or not with Pph. One-week post infection, the plants were collected and their CWs were extracted for the subsequent analyses.

## 2. Results

### 2.1. INA Reduced Bean Susceptibility to Pph and Increased Flg22-Triggered Response

To determine whether the INA application was able to prime bean plants defense as it was described before [42,49], the phenotypical symptoms after Pph inoculation were recorded. Plants were grown in vitro, a system in which infection can be forced through high pathogen concentration and humidity and without other environmental stresses [50,51]. When plants were 2 weeks old, the experimental conditions were established as follows: bean plants pre-treated with INA (INA) or not (Mock) were inoculated with Pph (INA + Pph, and Mock + Pph, respectively) (Supplementary Figure S1A). The observed disease symptoms were similar to those previously described [45]. Mock plants grew without symptoms (Figure 1A), as their foliar color and development corresponded with the normal growth of non-stressed plants. By contrast, Mock + Pph-inoculation triggered general chlorosis, greasy appearance on leaves, and the development of necrotic tissue areas, as shown in Figure 1B, indicating the susceptibility of these plants against the pathogen. INA did not cause visible stress symptoms compared to Mock (Figure 1C). Interestingly, foliar damage produced by the Pph inoculation was reduced considerably in INA + Pph plants (Figure 1D). In this case, chlorosis was restricted to foliar margins, and no greasy leaves appeared. Therefore, compared to Mock + Pph, INA + Pph plants were more resistant to Pph, and showed a statistically smaller lesion area (Figure 1E), and interestingly showed small necrotic spots in interveinal areas, which could be the result of a hypersensitive response (highlighted area in Figure 1D), as has been observed in other bean varieties resistant to Pph [52].

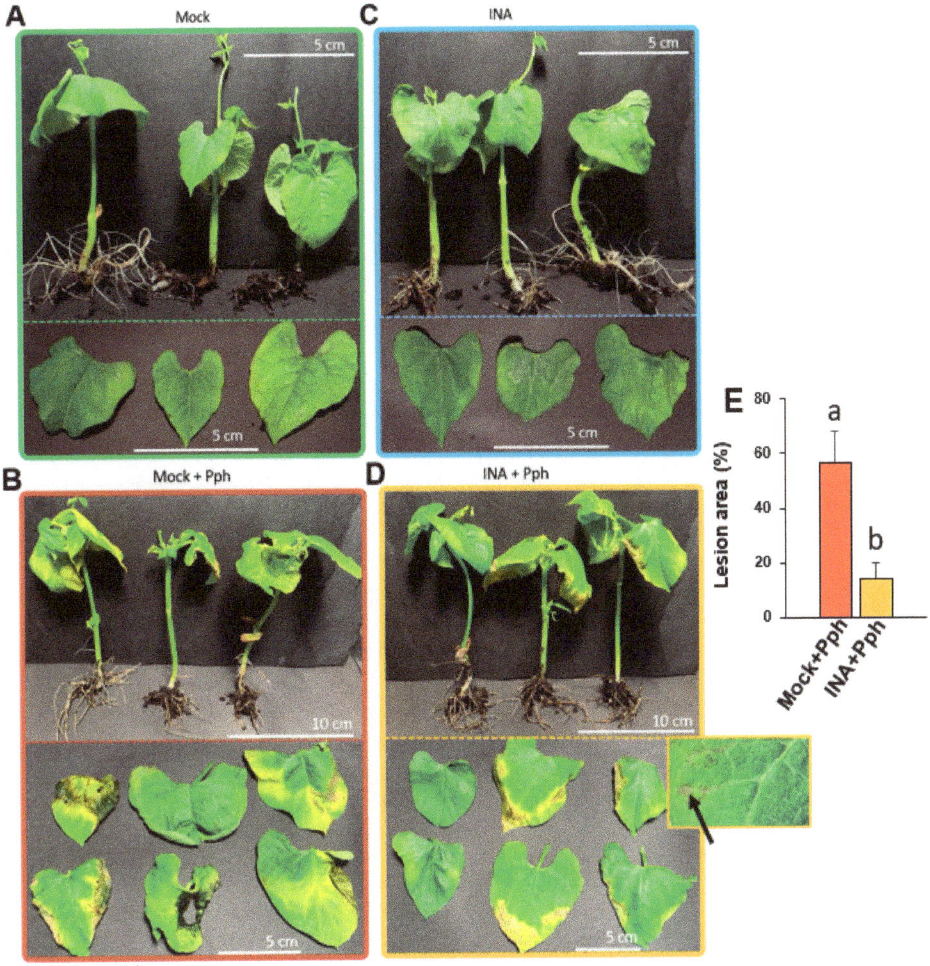


**Figure 1.** Phenotypic damage caused by the virulent bacteria *P. syringae* pv. *phaseolicola* (Pph) in common bean plants. Entire plants and extended leaves are shown to compare symptoms in Mock (**A**,**C**) and INA-pre-treated plants (**B**,**D**), non- (**A**,**B**) or Pph-inoculated (**C**,**D**). In panel D, a detail of the observed necrotic spots indicative of hypersensitive response (pointed by an arrow) is shown. (**E**) Quantification of the lesion area. Data represent mean ± SE ($n = 6$) and statistically differences, indicated by letters, were achieved according to *t*-Student test ($p < 0.05$). Pictures correspond to one experiment representative of three independent ones with similar results.

In view of the results obtained, whether INA priming was able to modify the ability of Riñón variety to trigger a general defense response was investigated. For this purpose, reactive oxygen species (ROS) production in bean leaf disks was monitored by a peroxidase/luminol-based assay (See Supplementary Figure S1B). This method is commonly used to reveal whether pathogen- or host-derived ligands are able to trigger early immune hallmarks [53]. In our experimental system, the INA application did not trigger a ROS production compared to mock (Figure 2). However, INA pre-treatment produced the highest ROS peak after the addition of the bacterial-derived peptide flg22 (which may mimic Pph inoculation). This suggested that INA did not induce a typical immune-associated fast response such as ROS production but primed the bean cells for more intense defense responses.

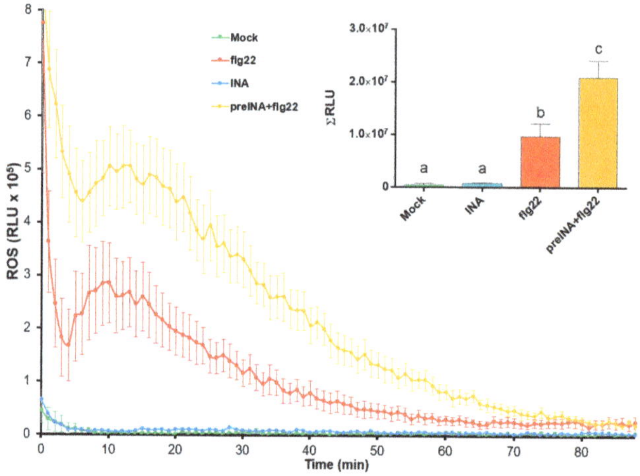


**Figure 2.** Reactive oxygen species (ROS) production in bean leaf-disks subjected to different treatments: water (Mock, green line), 100 μM INA (INA, blue line), 1 μM flg22 (flg22, red line), or 1 μM flg22 after pre-treatment with 100 μM of INA (preINA + flg22, yellow line), and measured as relative light units (RLU) produced by Luminol reaction over the time. Total areas under the curves were integrated, and resultant values are represented at the right side of the panel. Data represent mean ± SE ($n$ = 8) from one experiment representative of three independent ones with similar results. Statistically significant differences, indicated by letters, were achieved according to one-way ANOVA ($p < 0.05$), by post hoc Tukey test.

### 2.2. Pph Inoculation and INA Priming Produce Different CW Fingerprinting Than That Observed in the Mock

Several recent evidence support a more prominent role of the CW in plant immunity than previously believed, although the knowledge about plant CW remodeling after pathogen infection and/or immune priming is scarce [4,5,22,26]. With this in mind, CWs from Mock, Mock + Pph, INA and INA + Pph bean plants were extracted. Characterization of whole CWs began with the non-invasive technique FTIR-spectroscopy, used to obtain global fingerprints of the CW [54]. The FTIR spectra obtained were clustered in a Principal Component Analyses (PCA), which separated the treatments in two groups (Mock and the rest of the treatments) according to dimension 2, which explained 29.43% of variance. In order to predict the putative CW components contributing to such segregation, the wavenumbers whose values had a higher correlation with dimension 2 were extracted and summarized in Table 1 together with their associated CW components. Several wavenumbers were associated with RG-I, galactan and xyloglycan, (1148 and 1152 cm$^{-1}$), pectins (1232 and 1244 cm$^{-1}$), uronic acids (1616–1628 cm$^{-1}$), or with arabinogalactans (1156 cm$^{-1}$). In addition, the wavenumbers related to cellulose (1160 and 1164 cm$^{-1}$) and phenolics (1632 and 1720 cm$^{-1}$) were also found. Next, difference spectra against mock were calculated for each condition (Figure 3B). As a result, the profile between 1800 and 1170 cm$^{-1}$ approximately, which referred among others to wavenumbers regarding uronic acids and pectins, was similar among all conditions. By contrast, the region from 1170 to 800 cm$^{-1}$ associated to RG-I, Galactan, Xyloglycan, Arabinogalactan and Cellulose (Figure 3B, Table 1) only changed after INA pre-treatment (with or without further Pph infection) when compared to mock.

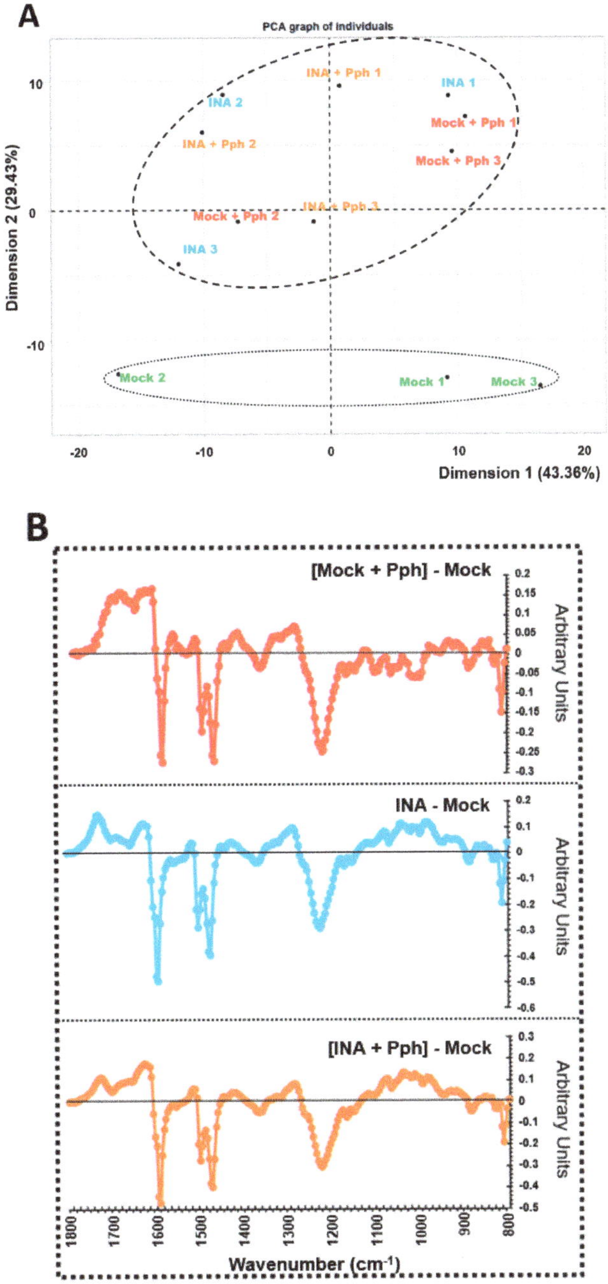


**Figure 3.** Cell wall fingerprinting. (**A**) Principal Component Analysis obtained from FTIR spectra of crude cell walls extracted from leaves of Mock, Mock + Pph, INA, and INA + Pph common bean plants. Each dot represents a biological replicate obtained in each of the 3 independent experiments carried out. PCA is plotted with dimensions 1 and 2, which explained the 43.36% and 29.43% of the total variance, respectively. (**B**) Average FTIR difference spectra obtained by digital subtraction of the Mock CW spectra from the CW spectra of the other treatments.

**Table 1.** FTIR wavenumbers ($cm^{-1}$) with highest contribution to explain Dimension 2 as Cos2 Dim2 (Figure 1) and the CW component to which they are associated.

| Wavenumber ($cm^{-1}$) | Cos2 Dim2 | CW Component | Reference |
|---|---|---|---|
| 808–816 | $8.05 \times 10^{-1}$–$6.93 \times 10^{-1}$ | Unknown | |
| 1148 | $7.14 \times 10^{-1}$ | RG-I, Galactan, Xyloglucan | [55] |
| 1152 | $7.98 \times 10^{-1}$ | RG-I, Galactan, Xyloglucan | [55] |
| 1156 | $7.18 \times 10^{-1}$ | Arabinogalactan | [55] |
| 1160 | $6.90 \times 10^{-1}$ | Cellulose | [56,57] |
| 1164 | $7.39 \times 10^{-1}$ | Cellulose | [58] |
| 1192–1196 | $6.43 \times 10^{-1}$–$7.26 \times 10^{-1}$ | Unknown | |
| 1200 | $7.61 \times 10^{-1}$ | C = CH | [59] |
| 1204–1228 | $7.98 \times 10^{-1}$–$8.12 \times 10^{-1}$ | Unknown | |
| 1232 | $8.02 \times 10^{-1}$ | Pectin | [60] |
| 1236–1240 | $7.72 \times 10^{-1}$–$7.26 \times 10^{-1}$ | Unknown | |
| 1244 | $6.67 \times 10^{-1}$ | Pectin | [59] |
| 1472–1596 | $6.61 \times 10^{-1}$–$6.14 \times 10^{-1}$ | Unknown | |
| 1616 | $7.34 \times 10^{-1}$ | Free carboxyl uronic acid | [55] |
| 1620 | $7.15 \times 10^{-1}$ | Free carboxyl uronic acid | [60] |
| 1624 | $6.89 \times 10^{-1}$ | Free carboxyl uronic acid | [60] |
| 1628 | $7.24 \times 10^{-1}$ | Free carboxyl uronic acid | [60] |
| 1632 | $7.41 \times 10^{-1}$ | Phenolic ring | [60] |
| 1636–1648 | $7.57 \times 10^{-1}$–$6.35 \times 10^{-1}$ | Unknown | |
| 1720 | $6.09 \times 10^{-1}$ | Phenolic ester | [57,60] |

### *2.3. INA Priming Induced Quantitative Changes in CW Polysaccharides Not Observed after the Pph Inoculation*

To deepen into the CW changes after the Pph inoculation and the INA priming, the cellulose content was measured in crude CWs. It should be taken into account that cellulose, a major CW component, has been described to participate in defense and remodeling [20]. Besides, our results indicate that cellulose contributes to discriminate among treatments after PCA of FTIR CW spectra (Table 1). Updegraff methodology confirmed that the INA pre-treatment resulted in an increased crystalline cellulose content in both INA and INA + Pph plants, which could indicate a CW reinforcement after priming, due to the fact that the simple Pph inoculation of Mock did not alter the cellulose content significantly (Figure 4).

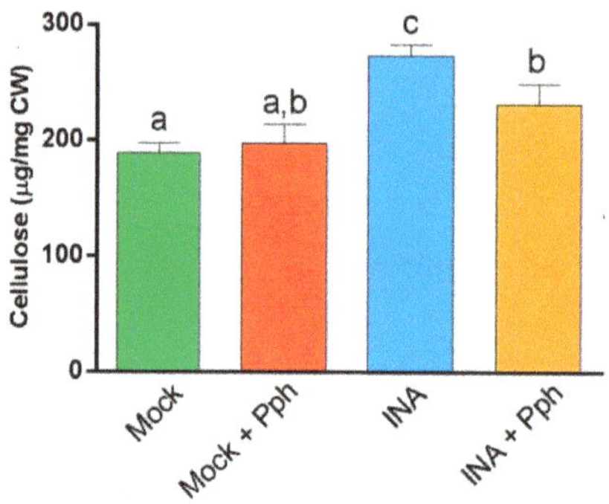


**Figure 4.** Cellulose content (µg per mg of dry weight CW) of the indicated treatments. Data represent mean ± SE ($n = 3$). Statistical analysis was carried out by one-way ANOVA where letters indicate differences by Tukey test ($p < 0.05$).

In order to explore the possibility of a CW reinforcement occurring after INA priming, a sequential CW extraction of matrix polysaccharides was performed. During the CW fractionation, pectic polysaccharides are commonly extracted in CDTA and $Na_2CO_3$ fractions, while hemicelluloses are mainly extracted by incubation with different alkali (normally

KOH or NaOH) concentrations, which extract hemicelluloses slightly (KI fraction) and tightly (KII fraction) bound to cellulose and/or cross-linked (Cosgrove, 2018). The following fractions recover polysaccharides tightly bound to cellulose (SnCR) and amorphous cellulose (TFA) [61], respectively.

The Pph inoculation only had an effect in pectic fractions, producing an increment of neutral sugars and uronic acids in CDTA fraction as well as an increment of uronic acids in $Na_2CO_3$ fraction compared to mock (Figure 5). However, the INA pretreatment produced dramatic changes in the neutral sugar distribution, by decreasing its content in CDTA, $Na_2CO_3$ and SnCR fractions, and significantly increasing it in TFA fraction (Figure 5A). The uronic acid content diminished in CDTA by INA pretreatment, but increased in pectic ($Na_2CO_3$) and hemicellulosic (KII) fractions, as well as in the SnCR fraction (Figure 5B). These results show a clear shift of polysaccharides from the CDTA fraction to others where extracted polymers were linked to the matrix more strongly, thus suggesting changes in their composition and/or structure. Interestingly, these CW changes observed in INA were similarly observed after the Pph inoculation (INA + Pph).

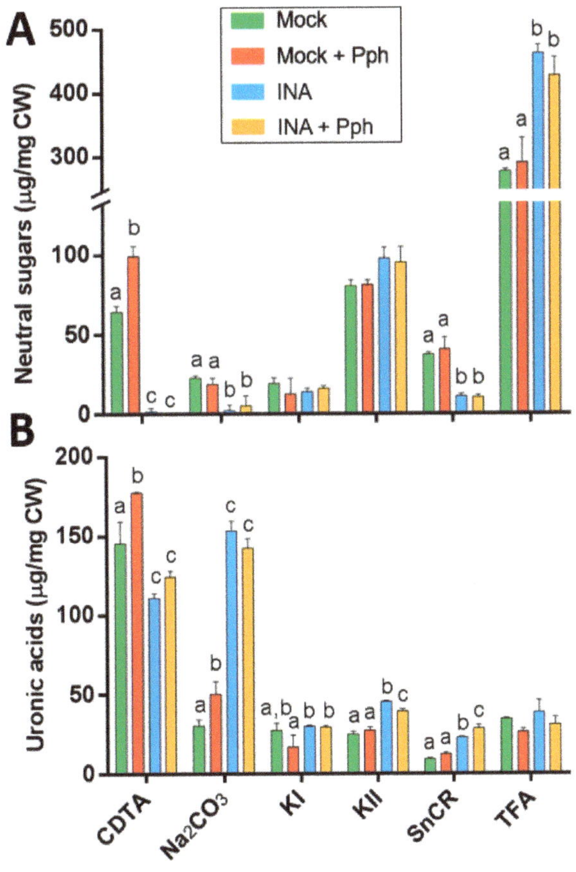


**Figure 5.** Biochemical composition of CW fractions from Mock, Mock + Pph, INA and INA + Pph inoculated common bean plants: (**A**) neutral sugars and (**B**) uronic acid content. Data represent mean ± SE ($n$ = 3). Statistical analysis was carried out by one-way ANOVA where letters indicate differences by Tukey test ($p < 0.05$).

Thereafter, the fractions were subjected to acid hydrolysis, and neutral sugar composition was determined by gas chromatography (Figure 6). After the INA pretreatment, no neutral sugars were detected in CDTA and $Na_2CO_3$ fractions, probably due to an increment in the uronic acid to neutral sugar proportion in these fractions (Figure 5), which makes the neutral sugar quantification difficult [62]. In these pectin-enriched fractions (Figure 6A,B), the Pph inoculation produced increments in rhamnose (Rha) and arabinose (Ara) concentrations, most likely arising from RG-I. Additionally, Ara and galactose (Gal) also increased in SnCR and TFA fractions in Mock + Pph (Figure 6E,F), pointing also to an enrichment in arabinogalactans.

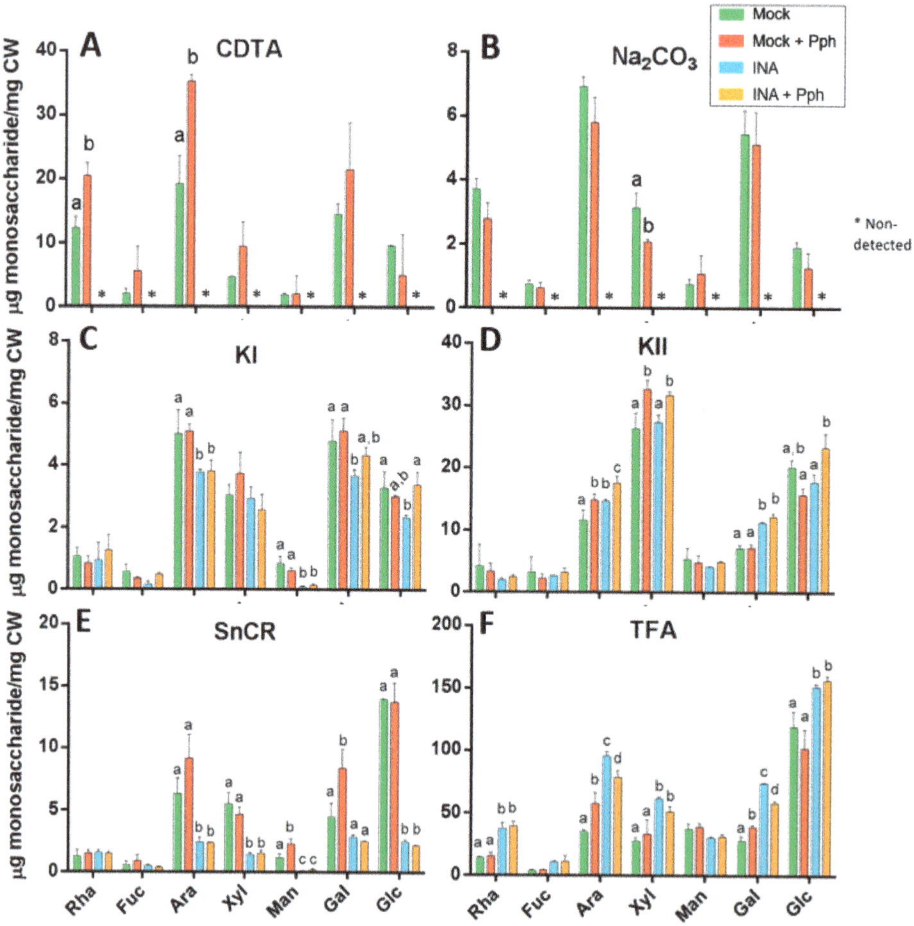


**Figure 6.** Monosaccharide analysis from different CW fractions: CDTA (**A**), $Na_2CO_3$ (**B**), KI (**C**), KII (**D**), SnCR (**E**), and TFA (**F**) of Mock, Mock + Pph, INA, and INA + Pph common bean plants. Monosaccharides are referred as Rha: rhamnose, Fuc: fucose, Ara: arabinose, Xyl: xylose, Man: mannose, Gal: galactose and Glc: glucose. Data represent mean ± SE ($n$ = 3). Statistical analysis was carried out by $t$-student for Figure A and B, and by one-way ANOVA for Figure C to F where letters indicate differences by Tukey test ($p < 0.05$). Asterisks refers to non-detected monosaccharides.

In the INA pretreatments a displacement of Rha, Ara, xylose (Xyl), Gal, and glucose (Glc) monosaccharides, from KI, KII, and SnCR fractions to the TFA fraction, was observed (Figure 6C–F), indicating that the polysaccharides implicated in the CW remodeling after priming could be RG, XG, and/or xylans, more tightly linked to the matrix. A decrease in the mannose (Man) content was observed in KI and SnCR fractions after the INA pretreatments compared with Mock. In addition, the Glc increase found in the TFA fraction after the INA pretreatment (Figure 6F) could be related with the higher cellulose content found in crude CWs for those conditions (Figure 4) or even to an increment in other beta-linked-glucans able to reinforce the CW structure such as callose [63,64].

Finally, when plants were inoculated with Pph after priming (INA + Pph), no substantial changes were observed compared with INA, except an increment in Gal and Glc in KI (Figure 6C), and in Ara, Xyl, and Glc in KII fractions (Figure 6D), which could derive from hemicelluloses

### 2.4. *INA Priming Increased the Average Molecular Weight of Polysaccharides in All CW Fractions*

Additional information about relative mass distribution and average molecular weights ($M_W$) of polysaccharides in the CW matrix was obtained by gel permeation chromatography [65] (Figure 7). Mock + Pph involved a $M_W$ increase of all fractions compared to Mock (Figure 7), except in KII, being the most notorious increase in the $Na_2CO_3$-pectic fraction (Figure 7B), where polysaccharide dispersion was higher. Interestingly, the INA pretreatment was always associated with a remarkable increase in the estimated $M_W$ of all polysaccharide populations analyzed. This shift was particularly noticeable in hemicellulosic fractions in which a 14-fold (KI fraction) and 6.5-fold increase (KII fractions) was found (Figure 7C,D). In these cases, INA + Pph had nearly negligible impact on $M_W$ compared to INA. These results obtained after INA priming pointed to the possibility of CW remodeling by an extensive polymer cross-linking. Interestingly, the elution profile shown two populations of polysaccharides after the INA pretreatment in fractions CDTA (Figure 7A), KI (Figure 7C) and KII (Figure 7D): (i) relative to lower $M_W$, which corresponded with Mock and Mock + Pph population, and (ii) of higher $M_W$, that did not appear in those not treated with INA.

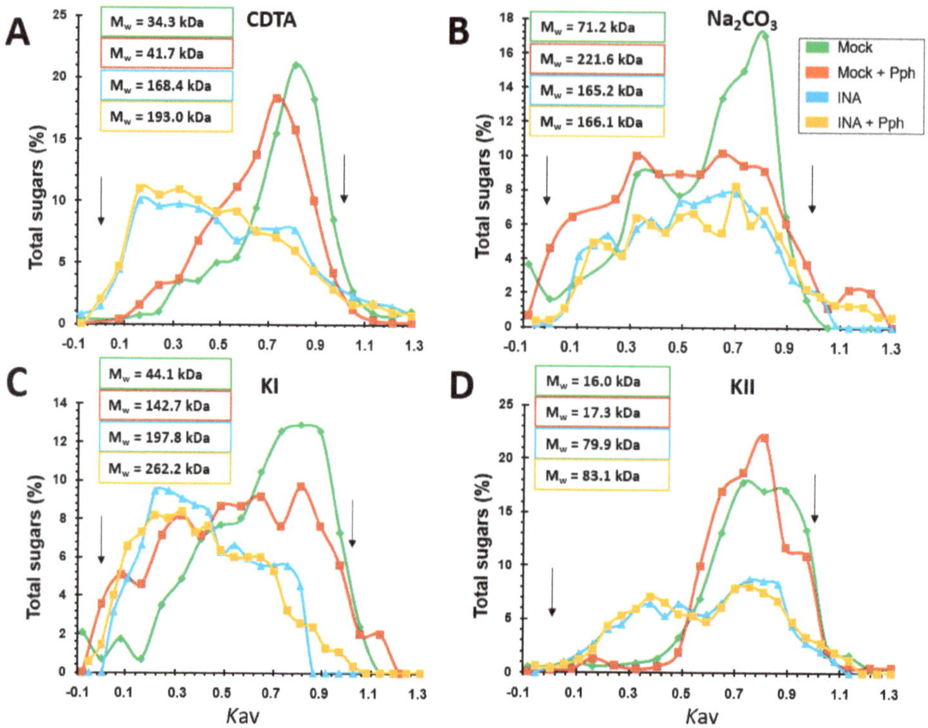


**Figure 7.** Relative mass distribution profiles of CDTA (**A**), $Na_2CO_3$ (**B**), KI (**C**), and KII (**D**) fractions extracted from CWs of Mock, Mock + Pph, INA and INA + Pph. 0 and 1 *Kav* were assigned to blue dextran and sucrose markers, respectively (indicated with arrows). The average molecular weight in kDa ($M_W$) obtained by *Kav*(1/2) method is shown inserted in boxes.

### 2.5. Qualitative Epitope Changes in CWs Produced by the Pph Inoculation and INA Priming

As the Pph inoculation, as well as INA priming, involved a displacement of polysaccharides populations extracted across all the fractions (Figures 5 and 6), and the increment in $M_W$ of these polysaccharides (Figure 7), an epitope characterization analysis of these CWs was carried out (Figure 8). Therefore, a screening of CW qualitative changes was performed by immunodot assays (IDAs) against different CW proteins such as extensins (recognized by LM1 monoclonal antibody [66]) and arabinogalactan proteins (AGPs, recognized by LM2 [9]). HGs with different degree and pattern of methyl esterification were screened by means of JIM5 or JIM7 antibodies, which probe to low and high methyl esterified HGs respectively [67]. The β-1,4-galactan and α-1,5-arabinan side chains of RG-I were also probed by using LM5 or LM6 antibodies, respectively [9]). In addition, hemicellulose epitopes were analyzed with LM11, which detects xylans and AXs [68]), and LM15, which labels non-fucosylated XG [69]).

The upper side of the heatmap shows the relationship among antibodies binding profiles. Xylan/arabinoxylan (LM11) and extensin (LM1) antibodies clustered together (cluster A) and had less CW epitopes than pectic and hemicellulosic antibodies, which clustered in the other group (cluster B). Within group B, HG antibodies (JIM5 and JIM7) had the more intense binding profile (cluster B.1), while RG-I (LM5 and LM6), XG (LM15), and AGP (LM2) antibodies grouped in an independent sub-cluster (B.2).

Attending to the left part of the heatmap, two clusters were observed: cluster C, which included different fractions from INA-pretreatments (INA and INA + Pph), and cluster D, in which $Na_2CO_3$, KI, and SnCR fractions from Mock and Mock + Pph plants were included. Interestingly, $Na_2CO_3$ and KI fraction always grouped together although in different clusters depending on the treatment (cluster C.1a for INA and INA + Pph, and cluster D for Mock and Mock + Pph), whereas CDTA and KII fractions were clustered independently in cluster C.1b and cluster C.2, respectively. The clustering analysis after IDAs revealed a similar epitope profile for CDTA and 4N KOH (KII) extracted polysaccharides independently of the treatment. Besides, INA and INA + Pph linked to a differentiated labelling pattern in $Na_2CO_3$ and 0.1N KOH (KI) extracted polysaccharides when compared to Mock or Mock + Pph. As expected, in Mock plants the pectin fraction extracted with CDTA was enriched in HG with high and low degree of methyl esterification (JIM7 and JIM5) and RG-I (LM5 and LM6), but also non-fucosylated XG and AGPs were found. Mock + Pph produced only a slight increase in HG methylation degree (JIM7) in the CDTA fraction. However, INA and INA + Pph also increased the presence of extensins (LM1) and AGPs (LM2), and decreased galactan side chains of RG-I (LM5) and XG (LM15) epitopes, mainly in INA + Pph.

In Mock, CW polysaccharides extracted in $Na_2CO_3$ and KI fractions were enriched in RG-I (LM5 and LM6), HG (JIM5 and JIM7), XG (LM15) and AGPs (LM2), but the $Na_2CO_3$ fraction had only a small proportion of extensin epitopes (LM1), and KI had only a small proportion of xylan epitopes (LM11). Mock + Pph did not substantially change the epitope distribution in $Na_2CO_3$ and KI fractions. However, the epitope profile of these fractions dramatically changed after INA pretreatment, mainly due to the increase in extensin (LM1) and HG (JIM5 and JIM7) epitopes, especially in the $Na_2CO_3$ fraction after the Pph inoculation (INA + Pph).

The KII fraction was mainly composed of XG (LM15) and xylans (LM11) in Mock, although HG (JIM5 and JIM7), RG-I (LM5 and LM6) and AGP (LM2) epitopes were also detected. Mock + Pph itself did not substantially affect this profile, but the INA pretreatment produced a pectin increment, mainly HG with a high degree of methylation (JIM7) and RG-I (LM5 and LM6) and an increase in extensin epitopes.

Finally, the SnCR fraction was more variable in composition depending on the treatment. In Mock, SnCR was composed of pectins (RG-I and HG), hemicelluloses (XG and Xylans), and AGP. However, in Mock + Pph a decrease in hemicellulosic (LM11 and LM15) and HG (JIM5 and JIM7) epitopes was observed. The composition of this fraction was also different between INA pretreatments. The INA pretreatment increased, as in other fractions, the presence of HG (JIM5 and JM7) and extensin (LM1) epitopes, but the Pph inoculation (INA + Pph) also increase the abundance of hemicellulose (LM15) and AGP (LM2) epitopes compared to Mock.

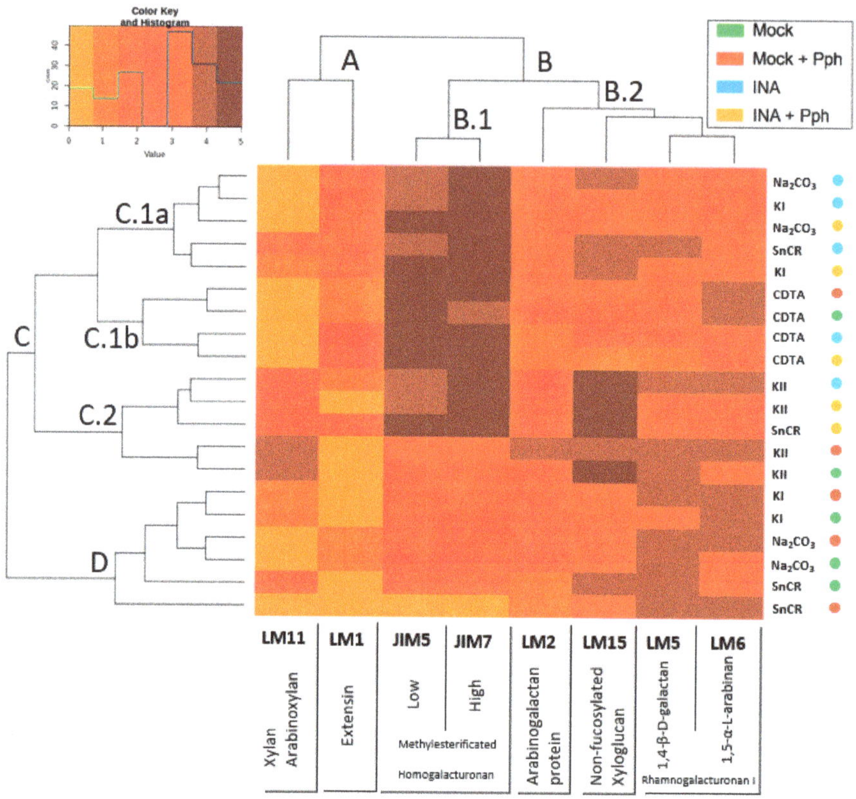


**Figure 8.** Heatmap of data obtained from IDAs of fractioned CW from Mock, Mock + Pph, INA and INA + Pph conditions. The value assigned from 0 to 5, depending on the marked dilution after development, is represented in a color key for the 5 levels. Fractions of each treatment are clustered in the horizontal axis, while monoclonal antibodies are clustered in the vertical axis. For easy visualization, the next clusters were established with respect to antibodies clusterization: cluster A, grouping LM1 and LM11; cluster B, divided into B.1, containing JIM5 and JIM7, and B.2, where LM5, LM6, LM2 and LM15 were included. Regarding fractions and treatments, the clusters obtained were: cluster C, divided into C.1a (relative to $Na_2CO_3$ and KI fraction of INA treatments), C.1b (where all CDTA fractions were located), and C.2, with KII fractions mostly; cluster D, relative to $Na_2CO_3$ and KI fraction but of Mock and Mock + Pph.

### *2.6. INA Priming Prevented Enzymatic Digestibility of CWs*

The CW changes observed after the Pph inoculation, and especially after the INA priming, demonstrated CW remodeling events. In order to know whether such CW remodeling confers more resistance to pathogen attack, especially taking into account that Pph is able to produce CWDE, a time course of sugar release after enzymatic hydrolysis of crude CWs was carried out (Figure 9). In this assay, no significant differences were found when both, sugar yield and sugar release kinetics obtained from Mock + Pph and Mock CWs were compared. However, INA pre-treatments induced a slower release of total sugars during the first half of the time course. Samples from INA + Pph did not reach sugar release levels of Mock after 24 h. In summary, the INA priming induced a CW reinforcement, making the structure more resistant to enzymatic hydrolysis.

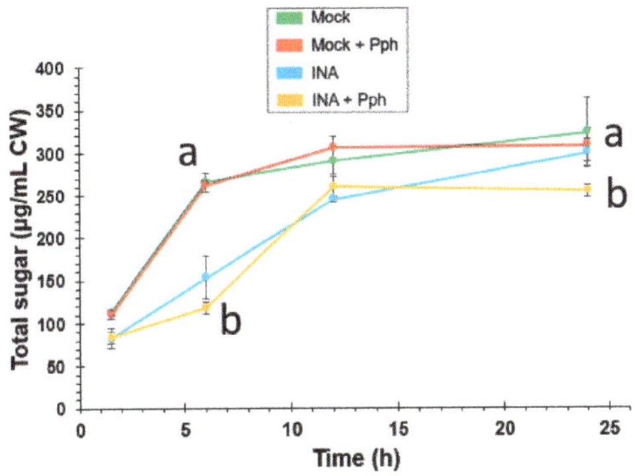


**Figure 9.** Time course of sugars released by enzymatic hydrolysis of CWs extracted from Mock, Mock + Pph, INA and INA + Pph conditions. Data represent mean ± SE ($n$ = 3). Statistical analysis was carried out by two-way ANOVA where letters indicate significant differences by Tukey test ($p < 0.05$).

## 3. Discussion

In nature, plants must overcome multiple stresses, such as those induced by pathogens, by activating several defense mechanisms. The CW is the first plant defensive layer against intruders. Plants are able to rearrange this structure on demand at different developmental stages. The CW can be reinforce by modifying the proportions of its main components, and/or by changing the type and extent of cross-linking among CW polymers [4]. To gain insight into the pathogen and priming capacity to reinforce the CW, the remodeling of CWs after the Pph inoculation and INA priming in the Pph-susceptible common bean variety *Riñón* was studied in this work.

Previous studies have suggested the ability of INA, as a SA structural analogue, to trigger the immune system in common bean [40,42,49]. To confirm the priming activation in *Riñón* common bean, plants of this variety previously treated or not with INA were inoculated or not with Pph (Supplementary Figure S1). By following that approach, it was confirmed that plants pre-treated with INA showed less halo blight symptoms [45], as the chlorosis and greasy appearance observed after the Pph inoculation decreased (Figure 1). The fact that these symptoms were constrained to damage at the foliar margins confirmed that INA pretreatment prior to the Pph inoculation protected *Riñón* bean plants from halo blight disease, as it was previously observed in other varieties [42,49]. As priming suggests, this observation may indicate that cell defensive responses have already been activated, and consequently, a stronger defensive mechanism is expected [37,41]. In order to confirm such hypothesis, the ROS burst triggered by peptide flg22 [70] was evaluated in bean leaf disks previously incubated or not with INA. As shown in Figure 2, a more intense ROS signal after flg22 triggering was detected in those plant disks previously treated with INA, which was in line with the priming concept, as it primed the plant cells to develop a more vigorously defense response somehow [71–74]. Ramirez-Carrasco et al. (2017) [42] proposed that this SA analogue produces epigenomic changes in bean that can even be inherited [43,49].

The presence of a CW that shields plants from pathogen invasion is a common resistance mechanism to disease among all plant cells. Several recent works have demonstrated a more relevant role of the CW than initially expected mediating such resistance. Most of them have focused on the study of mutants with a great variety of CW alterations, which have resulted in differential resistance or susceptibility against plant pathogens [4,20]. Other works have studied the role of elicitation with conserved CW-derived molecules from the pathogen, or from the plant itself, which once perceived the pathogen and triggered different immune responses [75,76]. However, the knowledge about CW remodeling after immune priming or bacterium infection is scarce. Here, we studied whether Pph inoculation and/or INA priming in common bean could promote a CW remodeling which would end up in the enhanced protection against halo blight disease, as shown in Figure 1.

The study of CW remodeling in this work began with the non-invasive technique FTIR spectroscopy of crude CWs, which resulted in the discrimination of FTIR spectra of Mock + Pph, INA and INA + Pph samples from the Mock samples (Figure 3). Dimension 2 of the PCA applied, which explained almost 30% of variance, was mainly related to wavenumbers associated with pectins, cellulose, and phenols (Table 1). The subsequent analysis of the polysaccharide content revealed changes in the amount of cellulose (Figure 4), matrix polysaccharide distribution (Figure 5), monosaccharide composition (Figure 6), and changes in the inmunoprofile of pectins (Figure 8).

When bean plants were inoculated with Pph, no statistical differences were found in cellulose content (Figure 4), in contrast with previous findings from other pathosystems [3]. However, our results point to both quantitative and qualitative changes in pectins and hemicelluloses after the Pph inoculation (Figures 5 and 6). The total amount of pectins (CDTA and $Na_2CO_3$ extracted polysaccharides) increased upon the Pph inoculation (Figure 5). Indeed, the higher HG methylation degree observed, as revealed by immunolabeling in these fractions (Figure 8), would suggest a HG synthesis increase after the infection, as HG methyl-esterification only occurs in Golgi apparatus [77–79], and/or a decrease of pectin methyl esterase activity after the Pph infection, as this is altered in other pathosystems [27,80]. Besides, the Ara and Xyl variation in the KII fraction suggest an increment of AX in tightly cross-linked hemicelluloses. The $M_w$ of pectins (particularly those $Na_2CO_3$ extracted) and loosely crosslinked hemicelluloses (KI fraction) raised after the Pph inoculation. A putative increase in the length and/or number of arabinan and galactan side chains of RG-I was also reported.

Several studies show how plants modify their pectin methylation degree or substitution pattern in order to avoid microbe colonization [78,81]. It is well-described that HG demethylation, as it is necessary to establish links between HG chains through $Ca^{2+}$ bonds to form the denominated egg-box complexes [82], participate in biotic resistance [30,31]. This type of link could explain the $M_w$ increase observed for pectic fractions (Figure 7) [83,84]. These changes in pectic polymers could be not enough as defensive mechanisms, as it was observed phenotypically (Figure 1). In fact, the CW degradability profile was similar in Mock + Pph and Mock (Figure 9). Pph is characterized, as other pathogenic microbes, by the production of CWDE that are released in the apoplast, similar to those used in the CW degradability assay [4,48]. This could be an explanation to the symptoms observed caused by the disease.

The Glc increase in the TFA fraction may reflect the accumulation of non-crystalline cellulose after the INA treatment (Figure 4). INA, as structural SA analogue, could stimulate cellulose synthesis. This polysaccharide is precisely regulated by growth factors, and previous studies have suggested the relation between SA and cellulose accumulation [85].

Alternative explanations for the Glc increase in the TFA fraction would be the accumulation of callose and/or the enrichment in a XG population tightly bound to cellulose. Both explanations seem plausible as the protective role of callose against Pph attack [86] and the function of XG by strengthening CW structure [63,64] have been previously reported. The CW fractionation of plants pre-treated with INA showed a noticeable displacement of matrix polysaccharides among fractions.

Regarding Mock and Mock + Pph conditions, neutral sugars were barely detected in CDTA and $Na_2CO_3$, decreased in SnCR, and increased in TFA (Figure 5). Specifically, the displacement of monosaccharides to the TFA fraction would suggest that RG-I, XG and xylans are the polysaccharides involved (Figure 6). A similar displacement from CDTA to $Na_2CO_3$, KII and SnCR could have occurred regarding uronic acids (Figure 5B). These changes indicated that polysaccharides become strongly attached to the matrix. Moreover, marked changes in polysaccharide size (Figure 7) and inmunoprofile in epitopes for each fraction (Figure 8) would suggest INA-dependent changes in the CW structure. The increase in glycan $M_w$ observed for all fractions in INA treatments compared to

Mock (Figure 7) could be explained by changes in cross-linking detected in CW epitopes (Figure 8), which is interesting when the sugar content decreased in CDTA and did not change in KI and KII (Figure 5). Galactan or arabinan side chain substitutions in RG-I decreased in CDTA but increased in $Na_2CO_3$ after the INA pretreatments. Additionally, AGPs or XG diminished after the priming, but all fractions obtained from plants pre-treated with INA showed an increment in HG and extensin epitopes (Figure 8). As being part of the same molecule [87,88], the changes observed in RG-I could have indirectly rendered a HG more attached to the matrix and, consequently, the detection of HG in fractions such as KI and KII, in which the presence of pectins is not usually abundant [89,90].

The more distinctive fact of CW fractions from INA treatments compared with those obtained from Mock and Mock + Pph CWs, was the increment in extensin epitopes, which could participate in matrix remodeling, maintaining the plasticity. Extensins are hydroxyproline-rich glycoproteins which interact with other CW components such as pectins, participating in the architecture, the structural organization and the strengthening of the CW [91–93]. Therefore, these proteins are involved in polysaccharide cross-linking [92,93]. Interestingly, extensin overexpression produces a decrease in the susceptibility of Arabidopsis to *Pseudomonas syringae* DC3000 [94]. In line with this, Arabidopsis plants with a reduced wall-extensin content are less resistant to *Botrytis* [26]. In addition, it is well known that some extensin genes are induced in *Arabidopsis* and *Nicotiana* by SA [95–97], and INA, as a SA analogue, could produce a similar effect in bean.

Taken together, all these changes suggest a remodeling of the CW architecture, which was finally confirmed by the enzymatic digestion assays, as CWs extracted from plants treated with INA showed a higher resistance to enzymatic hydrolysis (Figure 9). These differences, which were especially noticeable at early times, could be explained, at least partially, by a CW reinforcement after priming. The inoculation of the pathogen after the INA priming meant slight CW modifications, which was also reflected in a similar CW degrading profile upon enzymatic digestion. As previously reported [98], the CW cross-linking and strengthening could hamper pathogen CWDEs activity and confer biotic resistance, which would explain the slight lesions produced by halo blight after the INA pre-treatment in this variety (Figure 1). INA pre-treated plants were more resistant to Pph, and showed small necrotic spots, which resembled those produced during the HR response (Figure 1D). In resistant plant species, SA accumulates in the infection site at early stages of the immune response, which leads to HR [99]. However, previous studies showed that *Riñón* variety was not able to produce the early SA peak required after the Pph attack to induce HR [47]. Therefore, INA could replace the role of SA by increasing the defense signal at early stages after Pph infection in this variety.

To summarize, the Pph inoculation modified the common bean CW to a much lower extent than INA-priming. It did not involve an increase in cellulose content, but increased pectins (HG and RG) in pectic fractions and hemicelluloses (AX) in KII fraction. The $M_w$ of polysaccharides increased most of all in $Na_2CO_3$ fraction, but the CW was as vulnerable to enzymatic hydrolysis as Mock. By contrast, INA priming triggered a drastic CW remodeling, by increasing the cellulose content, displacing the matrix polysaccharides among fractions, and increasing the $M_w$ of polysaccharides extracted in each fraction. This could be related to an increment in pectins (HG and RG) and extensins in all fractions, which supported a more extensive cross-linking that resulted in a CW more recalcitrant to enzymatic hydrolysis. The Pph inoculation after INA priming did not modify substantially this CW remodeling and the resultant CW was as resistant to enzymatic degradation as INA without inoculation. Therefore, INA-priming-phenotypes regarding more intense ROS production after flg22-elicitation and increased resistance against Pph infection, could be explained, at least partially, by the CW remodeling observed. Future work to unveil the specific links between CW remodeling and disease resistance will pave the way to design novel crop protection strategies based on such knowledge.

## 4. Materials and Methods

### 4.1. Plant Material

Seeds of common bean *P. vulgaris* L. cultivar *Riñón*, from the Protected Geographical Identification *La Bañeza-León* region (Spain), were sterilized with 70% (*v*/*v*) ethanol for 30 s and 0.4% (*w*/*v*) NaClO for 20 min, prior to washing with sterile water. Germination took place under in vitro conditions using glass jars (946 mL of volume) with universal sterilized substrate (Blumenerde, Gramoflor, made in Germany). Plants were grown in a growth chamber at 25 ± 2 °C with a 16 h photoperiod under a photon flux density of 45 ± 5 µmol $m^{-2}s^{-1}$ provided by daylight fluorescent tubes (TLD 36W/830, Philips), as indicated in De la Rubia et al. (2021) [47].

### 4.2. Bacterial Strain and Growth Conditions

*Pseudomonas syringae* van Hall 1902, CECT321 (Pph) was grown for two days at 30 °C on liquid King's B (KB) medium at 220 rpm. For infection experiments, a final concentration of $10^8$ CFU/mL Pph was used. The Pph solution for inoculation was prepared by removing the media by centrifugation and resuspending in the same volume of sterile water, as indicated in De la Rubia et al. (2021) [47].

### 4.3. Elicitation, Pathogen Inoculation, and Sample Preparation

Leaves from common bean plants at V1 stage (with two cotyledonary leaves expanded, but not the true leaves developed) were sprayed with 2 mL per leaf of 100 µM 2,6-dichloropyridine-4-carboxylic acid (INA, Alfa Aesar, LOT: 10160271) (INA) or sterile water (Mock), as described in Martínez-Aguilar et al. [49]. After 7 days, some plants previously treated or not with INA were sprayed with 2 mL of the Pph solution on foliage leaves, resulting in Mock + Pph and INA + Pph treatments. All plants were grown for 7 more days and then the foliage leaves of 10 plants per treatment were collected and homogenized with liquid nitrogen to form a pool for each condition. All pools were stored at −80 °C until their use. Three complete independent experiments ($n$ = 3) were performed (see Supplementary Figure S1A).

### 4.4. Reactive Oxygen Species Detection

The ROS production was determined in V1 plants leaf disks using the luminol assay [100] in a Multi-Detection Microplate Reader Synergy HT (BioTek). Leaf disks were transferred to a multi-well plate (see Supplementary Figure S1B), containing 200 µL water or 100 µM INA, and they were incubated overnight. Afterwards, the previous solution was removed, and 100 µL per well of a solution containing 20 µM luminol L-012 (Wako) and 100 µg/mL peroxidase from horseradish type VI.A (Sigma, P6782) were added, and incubated in the dark for 30 min. Then, the reactions were started by adding 100 µL of water (Mock condition) as negative control, 100 µL of 2 µM flg22 (flg22 condition) as positive control, 100 µL of 200 µM INA (INA condition), and 100 µL of 2 µM flg22 to those disks previously incubated overnight with 100 µM INA (preINA + flg22). The ROS production, measured as relative luminescent units (RLUs), was measured over 90 min, with an integration time of 0.6 s. Data shown represent mean ± SE from one representative experiment of three independent ones performed with similar results (see Supplementary Figure S1A). The total areas under the kinetic curves were integrated, and the resulting values were statistically analyzed by One way ANOVA ($p < 0.05$), post hoc Tukey test.

### 4.5. Cell Wall Isolation

The plant material (a pool of foliage leaves of 10 plants for each replicate) was powdered by homogenizing the samples in liquid nitrogen with a mortar and pestle. Three g of fresh weight powder were then treated with 50 mL 70% ethanol (*v*/*v*) for 24 h, centrifuged to remove the supernatant (×2), and then incubated with 50 mL 80% acetone (*v*/*v*) for 24 h (×2). The insoluble residues were incubated with 50 mL 2.5 U $mL^{-1}$ α-amylase (Sigma type VI-A) in 0.01 M phosphate buffer pH 7.0 for 24 h at 37 °C (×2). Then, the remaining pellets were treated with 50 mL phenol-acetic acid-water (2:1:1 *v/v/v*) for 16 h at room temperature, with a change of solvents after 8 h of incubation. Finally, 50 mL 70% ethanol (*v/v*) (×3) and 50 mL 100% acetone (*v/v*) (×3) were sequentially added to wash the samples, and the final air-dried residues were considered CWs [61].

### 4.6. FTIR Spectroscopy

Crude CWs were used to obtain the FTIR spectra using a Jasco 4700 instrument (Tokyo, Japan). The average FTIR spectra ($n$ = 10), from 800 to 1800 $cm^{-1}$, were normalized and baseline-corrected with an ATR module of 4 $cm^{-1}$ resolution, using Spectra Manager version 2 (2016) software by Jasco corporation (Tokyo, Japan). A Principal Component Analysis was carried out from normalized spectra with R [101], and visualized with the FactoMineR package [102].

### 4.7. Sequential Polysaccharide Extraction

The sequential polysaccharide extraction followed the protocol previously described by Rebaque et al. (2017) [61] with slight modifications. Crude CWs (10 mg CW/mL) were treated with cyclohexane-trans-1,2-diamine-*N,N,N′,N′*,-tetraacetic acid sodium salt (CDTA) at pH 6.5 for 8 h. After centrifugation, the pellets were washed with distilled water. The residues obtained were treated with 0.05 M $Na_2CO_3$ + 0.02 M $NaBH_4$ and washed with distilled water, to obtain the $Na_2CO_3$ fractions. The residues obtained after centrifugation were then incubated in 0.1 N KOH + 20 mM $NaBH_4$ for 8 h, and washed with distilled water, obtaining the KI fractions. For the KII fractions, the remaining pellets were treated with 4 N KOH + 20 mM $NaBH_4$ for 8 h and washed with distilled water. KI and KII fractions were acidified to pH 5.0 with pure glacial acetic acid, and after that, the samples were agitated and centrifugated. Water was added to the pellets obtained after KII fraction, and then were acidified to pH 5 by adding pure glacial acetic acid. The extracts and their respective washings were combined. After its centrifugation, the supernatants were collected to form the Supernatant-Cellulose Residue (SnCR) fractions. The remaining residues were hydrolyzed with 3 mL 2N trifluoroacetic acid (TFA) for 2.5 h at 120 °C, centrifuged, and clarified supernatant was referred as TFA fraction.

### 4.8. Cell Wall Sugar Content Analysis

The cellulose content of CWs was quantified by the Updegraff method [103], under the hydrolytic conditions described by Saeman, [104], using glucose as standard.

The total sugars and uronic acids were determined by the phenol-sulfuric acid method [105], and the m-hydroxydiphenyl method [106], using D-glucose and galacturonic acid as reference, respectively. For the neutral sugar estimation, the values for total sugars and uronic acids were subtracted.

For the neutral sugar composition, samples from each fraction were hydrolyzed with 2 N TFA for 1 h at 121 °C, which resulted in monosaccharides that were derivatized to alditol acetates following the method described by Albersheim [107]. Furthermore, the alditol acetates were quantified by gas chromatography (GC) using a Perkin-Elmer equipment with a flame ionization detector (GC-FID), using a Supelco SP-2330 column and a Perkin-Elmer GC-FID, as described in Rebaque et al. (2017) [61]. Inositol was used as internal control, and monosaccharides L(−)rhamnose (Merck), L(−)fucose (Sigma), L(+)arabinose (Merck), D(+)xylose (Merck), D(+)mannose (Merck), D(+)galactose (Merck), and D(+)glucose (Panreac) as standard markers.

### 4.9. Gel Permeation Chromatography

The CW polysaccharides were size-fractionated by using a sepharose CL-4B column in a 120 mL bed-volume (1.5 cm diameter) column in pyridine/acetic acid/water (1/1/23 *v/v/v*) at 0.3 mL/min. The column was calibrated with sucrose and different commercial dextrans of known relative average molecular weight ($M_w$) as described by Kerr and Fry [65]. By using the $K_{av(1/2)}$ method [65], a calibration curve was obtained (log $M_w$ =

$-5.333\ K_{av(1/2)} + 8.103$). $V_0$ and $V_i$ ($K_{av}$ 0 and 1) were defined by dextran blue (2000 kDa) and sucrose, respectively. Then, the polysaccharides were loaded into the column at a concentration of 100 µg/mL, and the total sugars were estimated for each fraction. The nominal $M_w$ for each fraction was determined (nominal rather than absolute due to the conformational differences between dextran standards and CW polysaccharides).

### 4.10. Immunodot Assays

IDAs were carried out as described by García-Angulo et al. (2006) [89]. The reference compounds were commercial pectin (P41) and arabic gum. Monoclonal antibodies (mAbs) LM1, LM2, JIM5, JIM7, LM5, LM6, LM11, and LM15 were used with mung bean as standard. The reference compounds, as well as samples of fractions, were diluted 1/5, five times. Then aliquots of 1 µL from each dilution were spotted on nitrocellulose membranes. After drying, the membranes were blocked with 0.14 M NaCl, 2.7 mM KCl, 7.8° mM $Na_2HPO_4.12H_2O$ and 1.5° mM $KH_2PO_4$, pH 7.2 (PBS), with 4% fat-free milk powder, during 1.5 h at room temperature and incubated in primary antibody (hybridoma supernatants diluted 1/10). After washing the membranes, they were incubated in secondary antibody (antirat horseradish peroxidase conjugate, Sigma) diluted 1/1000, during 1.5 h at room temperature. The color was developed with 25 mL deionized water, 5 mL methanol with 10 mg $mL^{-1}$ 4–chloro–1–naphtol, 30 µL 6% ($v/v$) $H_2O_2$, and stopped by washing the membranes. A value ranging from 0 to 5 was assigned depending on the number of colored spots that were shown after developing the membranes as described in De Castro et al. (2014) [108]. As it corresponded with dilutions, this was used to establish a semiquantitative scale which was processed in a Heatmap, and clustering was performed with R [101].

### 4.11. Cell Wall Degradability

The CW degradability was assayed on 5 mg of crude CWs hydrolyzed with a cocktail of hydrolytic enzymes, containing 1% macerozyme (R10, from *Rhizopus* sp., Duchefa, EC number 232-885-6), 1% driselase (from *Basidiomicetes*, Sigma, EC number 286-055-3), 1% cellulase (Onozuka R10, from *Trichoderma viride*, Phytotechnology Laboratories, EC number 3.2.1.4), and 1% endo-polygalacturonase (M2, from *Aspergillus aculeatus*, Megazyme, EC number 3.2.1.15), in 20 mM sodium acetate (pH 4.8) (Fornalé et al., 2012). Samples were incubated at 37 °C by shaking. Aliquots of the CW hydrolysate were collected at different times, and 10 µL of 99% formic acid (Panreac) was added to stop the enzymatic digestion. The total sugars were quantified by the phenol-sulfuric method [105].

### 4.12. Statistical Analyses and Software

Results were represented using GraphPad Prism 6, GraphPad Software (La Jolla, California USA; www.graphpad.com), while the statistical analyses were performed with SPSS software (IBM Corp. Released 2017. IBM SPSS Statistics for Windows, Version 25.0. Armonk, NY: IBM Corp). Data normality was checked firstly by Kolmogorov-Smirnov test. The ROS and CW polysaccharide and monosaccharide quantification were analyzed by one-way ANOVA with a Tukey post-test to evaluate the differences between treatments. Data on digestibility were analyzed by a two-way ANOVA with a Tukey post-test. Results were significantly different considering $p < 0.05$. As mentioned above, the PCA and Heatmap were analyzed with R [101].

Foliar lesion areas were quantified using imageJ software [109].

**Supplementary Materials:** The following are available online at https://www.mdpi.com/article/10.3390/plants10081514/s1, Figure S1: Experimental design. (A) Seeds of common bean P. vulgaris L. cultivar Riñón were grown in vitro. Leaves from common bean plants at V1 stage were sprayed with 2 mL of 100 µM INA (INA condition) per leaf or with sterilized water (Mock condition). After 7 days, some plants previously treated or not with INA were sprayed with 2 mL of the Pph solution on foliage leaves, resulting in INA + Pph and Mock + Pph treatments. All plants were grown for 7 days more and then, the foliage leaves of 10 plants per treatment were collected. The experiment was repeated three independent times ($n = 3$). Afterwards, CWs were isolated, fractionated and their components analyzed by means of different techniques. (B) The ROS production was determined in V1 plants leaf disks. The reactions were triggered by adding 100 µL water (Mock condition) as negative control, 100 µL 200 µM INA (INA condition) as negative control for toxicity, 100 µL 2 µM flg22 (flg22 condition) as positive control, and also 100 µL 2 µM flg22 to those disks previously incubated overnight with 100 µM INA (preINA+flg22) as experimental condition. The experiment, with 8 plants assayed, was repeated three independent times ($n = 3$).

**Author Contributions:** Conceptualization, A.G.D.l.R. and P.G.-A.; methodology, A.G.D.l.R., H.M., A.E., M.L.C. and P.G.-A.; software, A.G.D.l.R.; validation, A.G.D.l.R. and H.M.; formal analysis, A.G.D.l.R.; investigation, A.G.D.l.R. and P.G.-A.; data curation, A.G.D.l.R.; writing—original draft preparation, A.G.D.l.R. and P.G.-A.; writing—review and editing, A.G.D.l.R., H.M., M.L.C., A.E. and P.G.-A.; visualization, A.G.D.l.R. and P.G.-A.; supervision, H.M., A.E. and P.G.-A.; project administration, P.G.-A.; funding acquisition, P.G.-A. All authors have read and agreed to the published version of the manuscript.

**Funding:** This research was funded by the Spanish Ministry of Economy, Industry and Competitiveness, grant number RTC-2016-5816-2. The APC was funded by a PhD student grant from the Spanish Education Ministry (FPU17/05849).

**Institutional Review Board Statement:** Not applicable.

**Informed Consent Statement:** Not applicable.

**Data Availability Statement:** Not applicable.

**Acknowledgments:** We thank Rafael Calvo for the assistance with the English manuscript correction.

**Conflicts of Interest:** The authors declare no conflict of interest.

## References


1. Dangl, J.L.; Horvath, D.M.; Staskawicz, B.J. Pivoting the plant immune system from dissection to deployment. *Science* **2013**, *341*, 746–751. [CrossRef]
2. van Butselaar, T.; den Ackerveken, G. Salicylic Acid Steers the Growth-Immunity Tradeoff. *Trends Plant Sci.* **2020**, *25*, 566–576. [CrossRef]
3. Hématy, K.; Cherk, C.; Somerville, S. Host-pathogen warfare at the plant cell wall. *Curr. Opin. Plant Biol.* **2009**, *12*, 406–413. [CrossRef]
4. Bacete, L.; Mélida, H.; Miedes, E.; Molina, A. Plant cell wall-mediated immunity: Cell wall changes trigger disease resistance responses. *Plant J. Cell Mol. Biol.* **2018**, *93*, 614–636. [CrossRef]
5. Vaahtera, L.; Schulz, J.; Hamann, T. Cell wall integrity maintenance during plant development and interaction with the environment. *Nat. Plants* **2019**, *5*, 924–932. [CrossRef] [PubMed]
6. Zablackis, E.; Huang, J.; Müllerz, B.; Darvill, A.C.; Albersheim, P. Characterization of the Cell-Wall Polysaccharides of Arabidopsis Thaliana Leaves. *Plant Physiol.* **1995**, *107*, 1129–1138. [CrossRef] [PubMed]
7. Carpita, N.; McCann, M. The cell wall. *Biochem. Mol. Biol. Plants.* **2000**, 52–109.
8. Mohnen, D. Pectin structure and biosynthesis. *Curr. Opin. Plant Biol.* **2008**, *11*, 266–277. [CrossRef]
9. Smallwood, M.; Yates, E.A.; Willats, W.G.T.; Martin, H.; Knox, J.P. Immunochemical comparison of membrane-associated and secreted arabinogalactan-proteins in rice and carrot. *Planta* **1996**, *198*, 452–459. [CrossRef]
10. Cosgrove, D.J. Growth of the plant cell wall. *Nat. Rev. Mol. Cell Biol.* **2005**, *6*, 850–861. [CrossRef] [PubMed]
11. Silva, J.; Ferraz, R.; Dupree, P.; Showalter, A.M.; Coimbra, S. Three Decades of Advances in Arabinogalactan-Protein Biosynthesis. *Front. Plant Sci.* **2020**, *11*, 2014. [CrossRef]
12. Miedes, E.; Vanholme, R.; Boerjan, W.; Molina, A. The role of the secondary cell wall in plant resistance to pathogens. *Front. Plant Sci.* **2014**, *5*, 358. [CrossRef] [PubMed]
13. Vanholme, R.; De Meester, B.; Ralph, J.; Boerjan, W. Lignin biosynthesis and its integration into metabolism. *Curr. Opin. Biotechnol.* **2019**, *56*, 230–239. [CrossRef]
14. Paredez, A.R.; Somerville, C.R.; Ehrhardt, D.W. Visualization of cellulose synthase demonstrates functional association with microtubules. *Science* **2006**, *312*, 1491–1495. [CrossRef]
15. Zhu, C.; Ganguly, A.; Baskin, T.I.; McClosky, D.D.; Anderson, C.T.; Foster, C.; Meunier, K.A.; Okamoto, R.; Berg, H.; Dixit, R. The fragile Fiber1 kinesin contributes to cortical microtubule-mediated trafficking of cell wall components. *Plant Physiol.* **2015**, *167*, 780–792. [CrossRef]
16. Kim, S.J.; Brandizzi, F. The plant secretory pathway for the trafficking of cell wall polysaccharides and glycoproteins. *Glycobiology* **2016**, *26*, 940–949. [CrossRef]

17. Chormova, D.; Messenger, D.J.; Fry, S.C. Boron bridging of rhamnogalacturonan-II, monitored by gel electrophoresis, occurs during polysaccharide synthesis and secretion but not post-secretion. *Plant J.* **2014**, *77*, 534–546. [CrossRef] [PubMed]
18. Stratilová, B.; Kozmon, S.; Stratilová, E.; Hrmova, M. Plant Xyloglucan Xyloglucosyl Transferases and the Cell Wall Structure: Subtle but Significant. *Molecules* **2020**, *25*, 5619. [CrossRef] [PubMed]
19. Castilleux, R.; Plancot, B.; Vicré, M.; Nguema-Ona, E.; Driouich, A. Extensin, an underestimated key component of cell wall defence? *Ann. Bot.* **2021**, *127*. [CrossRef]
20. Molina, A.; Miedes, E.; Bacete, L.; Rodríguez, T.; Mélida, H.; Denancé, N.; Sánchez-Vallet, A.; Rivière, M.P.; López, G.; Freydier, A.; et al. Arabidopsis cell wall composition determines disease resistance specificity and fitness. *Proc. Natl. Acad. Sci. USA* **2021**, *118*. [CrossRef]
21. Ellis, C.; Karafyllidis, I.; Wasternack, C.; Turner, J.G. The Arabidopsis mutant cev1 links cell wall signaling to jasmonate and ethylene responses. *Plant Cell* **2002**, *14*, 1557–1566. [CrossRef] [PubMed]
22. Hernández-Blanco, C.; Feng, D.X.; Hu, J.; Sánchez-Vallet, A.; Deslandes, L.; Llorente, F.; Berrocal-Lobo, M.; Keller, H.; Barlet, X.; Sánchez-Rodríguez, C.; et al. Impairment of cellulose synthases required for Arabidopsis secondary cell wall formation enhances disease resistance. *Plant Cell* **2007**, *19*, 890–903. [CrossRef] [PubMed]
23. De Benedictis, M.; Brunetti, C.; Brauer, E.K.; Andreucci, A.; Popescu, S.C.; Commisso, M.; Guzzo, F.; Sofo, A.; Ruffini Castiglione, M.; Vatamaniuk, O.K.; et al. The Arabidopsis thaliana Knockout Mutant for Phytochelatin Synthase1 (cad1-3) Is Defective in Callose Deposition, Bacterial Pathogen Defense and Auxin Content, But Shows an Increased Stem Lignification. *Front. Plant Sci.* **2018**, *9*, 19. [CrossRef]
24. Hu, Q.; Xiao, S.; Wang, X.; Ao, C.; Zhang, X.; Zhu, L. GhWRKY1-like enhances cotton resistance to Verticillium dahliae via an increase in defense-induced lignification and S monolignol content. *Plant Sci.* **2021**, *305*. [CrossRef] [PubMed]
25. Gallego-Giraldo, L.; Liu, C.; Pose-Albacete, S.; Pattathil, S.; Peralta, A.G.; Young, J.; Westpheling, J.; Hahn, M.G.; Rao, X.; Paul Knox, J.; et al. Arabidopsis dehiscence zone polygalacturonase 1 (ADPG1) releases latent defense signals in stems with reduced lignin content. *Proc. Natl. Acad. Sci. USA* **2020**, *117*, 3281–3290. [CrossRef]
26. Reem, N.T.; Pogorelko, G.; Lionetti, V.; Chambers, L.; Held, M.A.; Bellincampi, D.; Zabotina, O.A. Decreased Polysaccharide Feruloylation Compromises Plant Cell Wall Integrity and Increases Susceptibility to Necrotrophic FungalPathogens. *Front. Plant Sci.* **2016**, *7*. [CrossRef]
27. Bethke, G.; Thao, A.; Xiong, G.; Li, B.; Soltis, N.E.; Hatsugai, N.; Hillmer, R.A.; Katagiri, F.; Kliebenstein, D.J.; Pauly, M.; et al. Pectin biosynthesis is critical for cell wall integrity and immunity in Arabidopsis thaliana. *Plant Cell* **2015**, *28*, 537–556. [CrossRef]
28. Bacete, L.; Mélida, H.; López, G.; Dabos, P.; Tremousaygue, D.; Denancé, N.; Miedes, E.; Bulone, V.; Goffner, D.; Molina, A. Arabidopsis response reGUlator 6 (ARR6) modulates plant cell-wall composition and disease resistance. *Mol. Plant-Microbe Interact.* **2020**, *33*, 767–780. [CrossRef]
29. Pogorelko, G.V.; Kambakam, S.; Nolan, T.; Foudree, A.; Zabotina, O.A.; Rodermel, S.R. Impaired chloroplast biogenesis in Immutans, an arabidopsis variegation mutant, modifies developmental programming, cell wall composition and resistance to Pseudomonas syringae. *PLoS ONE* **2016**, *11*. [CrossRef] [PubMed]
30. Corpo, D.; Del Fullone, M.R.; Miele, R.; Lafond, M.; Pontiggia, D.; Grisel, S.; Kieffer-Jaquinod, S.; Giardina, T.; Bellincampi, D.; Lionetti, V. AtPME17 is a functional Arabidopsis thaliana pectin methylesterase regulated by its PRO region that triggers PME activity in the resistance to Botrytis cinerea. *Mol. Plant Pathol.* **2020**, *21*, 1620–1633. [CrossRef] [PubMed]
31. Lionetti, V.; Fabri, E.; De Caroli, M.; Hansen, A.R.; Willats, W.G.T.; Piro, G.; Bellincampi, D. Three pectin methylesterase inhibitors protect cell wall integrity for arabidopsis immunity to Botrytis. *Plant Physiol.* **2017**, *173*, 1844–1863. [CrossRef]
32. Lionetti, V.; Giancaspro, A.; Fabri, E.; Giove, S.L.; Reem, N.; Zabotina, O.A.; Blanco, A.; Gadaleta, A.; Bellincampi, D. Cell wall traits as potential resources to improve resistance of durum wheat against Fusarium graminearum. *BMC Plant Biol.* **2015**, *15*, 1–15. [CrossRef]
33. Ribot, C.; Hirsch, J.; Balzergue, S.; Tharreau, D.; Nottéghem, J.L.; Lebrun, M.H.; Morel, J.B. Susceptibility of rice to the blast fungus, Magnaporthe grisea. *J. Plant Physiol.* **2008**, *165*, 114–124. [CrossRef] [PubMed]
34. Liu, G.-T.; Wang, B.-B.; Lecourieux, D.; Li, M.-J.; Liu, M.-B.; Liu, R.-Q.; Shang, B.-X.; Yin, X.; Wang, L.-J.; Lecourieux, F.; et al. Proteomic analysis of early-stage incompatible and compatible interactions between grapevine and P. viticola. *Hortic. Res.* **2021**, *8*, 100. [CrossRef]
35. von Tiedemann, A.; Koopmann, B.; Hoech, K. Lignin composition and timing of cell wall lignification are involved in Brassica napus resistance to Sclerotinia sclerotiorum stem rot. *Phytopathology* **2021**. [CrossRef] [PubMed]
36. Macharia, T.N.; Bellieny-Rabelo, D.; Moleleki, L.N. Transcriptome profiling of potato (*Solanum tuberosum* L.) responses to root-knot nematode (meloidogyne javanica) infestation during a compatible interaction. *Microorganisms* **2020**, *8*, 1443. [CrossRef]
37. Mauch-Mani, B.; Baccelli, I.; Luna, E.; Flors, V. Defense Priming: An Adaptive Part of Induced Resistance. *Annu. Rev. Plant Biol.* **2017**, *68*, 485–512. [CrossRef] [PubMed]
38. Kuć, J. Induced Immunity to Plant Disease. *Bioscience* **1982**, *32*, 854–860. [CrossRef]
39. Kauss, H.; Theisinger-Hinkel, E.; Mindermann, R.; Conrath, U. Dichloroisonicotinic and salicylic acid, inducers of systemic acquired resistance, enhance fungal elicitor responses in parsley cells. *Plant J.* **1992**, *2*, 655–660. [CrossRef]
40. Kauss, H.; Franke, R.; Krause, K.; Conrath, U.; Jeblick, W.; Grimmig, B.; Matern, U. Conditioning of Parsley (*Petroselinum crispum* L.) Suspension Cells Increases Elicitor-Induced Incorporation of Cell Wall Phenolics. *Plant Physiol.* **1993**, *102*, 459–466. [CrossRef]

41. Buswell, W.; Schwarzenbacher, R.E.; Luna, E.; Sellwood, M.; Chen, B.; Flors, V.; Pétriacq, P.; Ton, J. Chemical priming of immunity without costs to plant growth. *New Phytol.* **2018**, *218*, 1205–1216. [CrossRef] [PubMed]
42. Ramírez-Carrasco, G.; Martínez-Aguilar, K.; Alvarez-Venegas, R. Transgenerational Defense Priming for Crop Protection against Plant Pathogens: A Hypothesis. *Front. Plant Sci.* **2017**, *8*, 696. [CrossRef]
43. Martínez-Aguilar, K.; Hernández-Chávez, J.L.; Alvarez-Venegas, R. Priming of seeds with INA and its transgenerational effect in common bean (*Phaseolus vulgaris* L.) plants. *Plant Sci.* **2021**, *305*. [CrossRef]
44. Monteagudo, A.B.; Rodiño, A.P.; Lema, M.; De La Fuente, M.; Santalla, M.; De Ron, A.M.; Singh, S.P. Resistance to infection by fungal, bacterial, and viral pathogens in a common bean core collection from the Iberian Peninsula. *HortScience* **2006**, *41*, 319–322. [CrossRef]
45. Arnold, D.L.; Lovell, H.C.; Jackson, R.W.; Mansfield, J.W. Pseudomonas syringae pv. phaseolicola: From "has bean" to supermodel. *Mol. Plant Pathol.* **2011**, *12*, 617–627. [CrossRef] [PubMed]
46. González, A.M.; Godoy, L.; Santalla, M. Dissection of Resistance Genes to Pseudomonas syringae pv. phaseolicola in UI3 Common Bean Cultivar. *Int. J. Mol. Sci.* **2017**, *18*, 2503. [CrossRef]
47. De la Rubia, A.G.; Centeno, M.L.; Moreno-González, V.; De Castro, M.; García-Angulo, P. Perception and first defense responses against *Pseudomonas syringae* pv. *phaseolicola* in *Phaseolus vulgaris* L.: Identification of Wall-Associated Kinase receptors. *Phytopathology* **2021**. [CrossRef] [PubMed]
48. Fogliano, V.; Ballio, A.; Gallo, M.; Woo, S.; Scala, F.; Lorito, M. Pseudomonas lipodepsipeptides and fungal cell wall-degrading enzymes act synergistically in biological control. *Mol. Plant-Microbe Interact. MPMI* **2002**, *15*, 323–333. [CrossRef]
49. Martínez-Aguilar, K.; Ramírez-Carrasco, G.; Hernández-Chávez, J.L.; Barraza, A.; Alvarez-Venegas, R. Use of BABA and INA As Activators of a Primed State in the Common Bean (*Phaseolus vulgaris* L.). *Front. Plant Sci.* **2016**, *7*, 653. [CrossRef] [PubMed]
50. Kortekamp, A.; Zyprian, E. Characterization of Plasmopara-Resistance in grapevine using in vitro plants. *J. Plant Physiol.* **2003**, *160*, 1393–1400. [CrossRef]
51. Zebrowska, J.I. In vitro selection in resistance breeding of strawberry (Fragaria x ananassa duch.). *Commun. Agric. Appl. Biol. Sci.* **2010**, *75*, 699–704. [PubMed]
52. Tsiamis, G.; Mansfield, J.W.; Hockenhull, R.; Jackson, R.W.; Sesma, A.; Athanassopoulos, E.; Bennett, M.A.; Stevens, C.; Vivian, A.; Taylor, J.D.; et al. Cultivar-specific avirulence and virulence functions assigned to avrPphF in *Pseudomonas syringae* pv. phaseolicola, the cause of bean halo-blight disease. *EMBO J.* **2000**, *19*, 3204–3214. [CrossRef]
53. Rebaque, D.; del Hierro, I.; López, G.; Bacete, L.; Vilaplana, F.; Dallabernardina, P.; Pfrengle, F.; Jordá, L.; Sánchez-Vallet, A.; Pérez, R.; et al. Cell wall-derived mixed-linked β-1,3/1,4-glucans trigger immune responses and disease resistance in plants. *Plant J.* **2021**, *106*, 601–615. [CrossRef]
54. Largo-Gosens, A.; Hernández-Altamirano, M.; García-Calvo, L.; Alonso-Simón, A.; Álvarez, J.; Acebes, J.L. Fourier transform mid infrared spectroscopy applications for monitoring the structural plasticity of plant cell walls. *Front. Plant Sci.* **2014**, *5*, 303. [CrossRef]
55. Kačuráková, M.; Capek, P.; Sasinková, V.; Wellner, N.; Ebringerová, A. FT-IR study of plant cell wall model compounds: Pectic polysaccharides and hemicelluloses. *Carbohydr. Polym.* **2000**, *43*, 195–203. [CrossRef]
56. Wilson, R.H.; Smith, A.C.; Kacuráková, M.; Saunders, P.K.; Wellner, N.; Waldron, K.W. The mechanical properties and molecular dynamics of plant cell wall polysaccharides studied by Fourier-transform infrared spectroscopy. *Plant Physiol.* **2000**, *124*, 397–405. [CrossRef] [PubMed]
57. Carpita, N.C.; Defernez, M.; Findlay, K.; Wells, B.; Shoue, D.A.; Catchpole, G.; Wilson, R.H.; McCann, M.C. Cell Wall Architecture of the Elongating Maize Coleoptile. *Plant Physiol.* **2001**, *127*, 551–565. [CrossRef]
58. Hori, R.; Sugiyama, J. A combined FT-IR microscopy and principal component analysis on softwood cell walls. *Carbohydr. Polym.* **2003**, *52*, 449–453. [CrossRef]
59. Sekkal, M.; Dincq, V.; Legrand, P.; Huvenne, J.P. Investigation of the glycosidic linkages in several oligosaccharides using FT-IR and FT Raman spectroscopies. *J. Mol. Struct.* **1995**, *349*, 349–352. [CrossRef]
60. Sene, C.; McCann, M.C.; Wilson, R.H.; Grinter, R. Fourier-Transform Raman and Fourier-Transform Infrared Spectroscopy (An Investigation of Five Higher Plant Cell Walls and Their Components). *Plant Physiol.* **1994**, *106*, 1623–1631. [CrossRef]
61. Rebaque, D.; Martínez-Rubio, R.; Fornalé, S.; García-Angulo, P.; Alonso-Simón, A.; Álvarez, J.M.; Caparros-Ruiz, D.; Acebes, J.L.; Encina, A. Characterization of structural cell wall polysaccharides in cattail (*Typha latifolia*): Evaluation as potential biofuel feedstock. *Carbohydr. Polym.* **2017**, *175*, 679–688. [CrossRef] [PubMed]
62. Filisetti-Cozzi, T.M.C.C.; Carpita, N.C. Measurement of uronic acids without interference from neutral sugars. *Anal. Biochem.* **1991**, *197*, 157–162. [CrossRef]
63. Park, Y.B.; Cosgrove, D.J. Xyloglucan and its interactions with other components of the growing cell wall. *Plant Cell Physiol.* **2015**, *56*, 180–194. [CrossRef]
64. Singh, M.; Pahal, V.; Ahuja, D. Isolation and characterization of microfibrillated cellulose and nanofibrillated cellulose with "biomechanical hotspots". *Carbohydr. Polym.* **2020**, *234*. [CrossRef]
65. Kerr, E.M.; Fry, S.C. Pre-formed xyloglucans and xylans increase in molecular weight in three distinct compartments of a maize cell-suspension culture. *Planta* **2003**, *217*, 327–339. [CrossRef]
66. Smallwood, M.; Martin, H.; Knox, J.P. An epitope of rice threonine- and hydroxyproline-rich glycoprotein is common to cell wall and hydrophobic plasma-membrane glycoproteins. *Planta* **1995**, *196*, 510–522. [CrossRef]

67. Knox, J.P.; Linstead, P.J.; King, J.; Cooper, C.; Roberts, K. Pectin esterification is spatially regulated both within cell walls and between developing tissues of root apices. *Planta* **1990**, *181*, 512–521. [CrossRef]
68. Jones, L.; Seymour, G.B.; Knox, J.P. Localization of Pectic Galactan in Tomato Cell Walls Using a Monoclonal Antibody Specific to (1 [->] 4)-β-D-Galactan. *Plant Physiol.* **1997**, *113*, 1405–1412. [CrossRef] [PubMed]
69. Marcus, S.E.; Verhertbruggen, Y.; Hervé, C.; Ordaz-Ortiz, J.J.; Farkas, V.; Pedersen, H.L.; Willats, W.G.T.; Knox, J.P. Pectic homogalacturonan masks abundant sets of xyloglucan epitopes in plant cell walls. *BMC Plant Biol.* **2008**, *8*, 60. [CrossRef] [PubMed]
70. Zipfel, C.; Robatzek, S.; Navarro, L.; Oakeley, E.J.; Jones, J.D.G.; Felix, G.; Boller, T. Bacterial disease resistance in Arabidopsis through flagellin perception. *Nature* **2004**, *428*, 764–767. [CrossRef] [PubMed]
71. Espinas, N.A.; Saze, H.; Saijo, Y. Epigenetic control of defense signaling and priming in plants. *Front. Plant Sci.* **2016**, *7*, 1201. [CrossRef] [PubMed]
72. Ramirez-Prado, J.S.; Abulfaraj, A.A.; Rayapuram, N.; Benhamed, M.; Hirt, H. Plant Immunity: From Signaling to Epigenetic Control of Defense. *Trends Plant Sci.* **2018**, *23*, 833–844. [CrossRef]
73. Gong, B.Q.; Guo, J.; Zhang, N.; Yao, X.; Wang, H.B.; Li, J.F. Cross-Microbial Protection via Priming a Conserved Immune Co-Receptor through Juxtamembrane Phosphorylation in Plants. *Cell Host Microbe* **2019**, *26*, 810–822.e7. [CrossRef] [PubMed]
74. Hake, K.; Romeis, T. Protein kinase-mediated signalling in priming: Immune signal initiation, propagation, and establishment of long-term pathogen resistance in plants. *Plant Cell Environ.* **2019**, *42*, 904–917. [CrossRef]
75. Mélida, H.; Bacete, L.; Ruprecht, C.; Rebaque, D.; del Hierro, I.; López, G.; Brunner, F.; Pfrengle, F.; Molina, A. Arabinoxylan-Oligosaccharides Act as Damage Associated Molecular Patterns in Plants Regulating Disease Resistance. *Front. Plant Sci.* **2020**, *11*, 1210. [CrossRef]
76. Mélida, H.; Sopeña-Torres, S.; Bacete, L.; Garrido-Arandia, M.; Jordá, L.; López, G.; Muñoz-Barrios, A.; Pacios, L.F.; Molina, A. Non-branched β-1,3-glucan oligosaccharides trigger immune responses in Arabidopsis. *Plant J.* **2018**, *93*, 34–49. [CrossRef] [PubMed]
77. Driouich, A.; Follet-Gueye, M.-L.; Bernard, S.; Kousar, S.; Chevalier, L.; Vicré-Gibouin, M.; Lerouxel, O. Golgi-mediated synthesis and secretion of matrix polysaccharides of the primary cell wall of higher plants. *Front. Plant Sci.* **2012**, *3*, 79. [CrossRef] [PubMed]
78. Lionetti, V.; Cervone, F.; Bellincampi, D. Methyl esterification of pectin plays a role during plant-pathogen interactions and affects plant resistance to diseases. *J. Plant Physiol.* **2012**, *169*, 1623–1630. [CrossRef] [PubMed]
79. Ruano, G.; Scheuring, D. Plant Cells under Attack: Unconventional Endomembrane Trafficking during Plant Defense. *Plants* **2020**, *9*, 389. [CrossRef]
80. Liu, N.; Sun, Y.; Pei, Y.; Zhang, X.; Wang, P.; Li, X.; Li, F.; Hou, Y. A pectin methylesterase inhibitor enhances resistance to verticillium wilt. *Plant Physiol.* **2018**, *176*, 2202–2220. [CrossRef]
81. Ferrari, S.; Sella, L.; Janni, M.; De Lorenzo, G.; Favaron, F.; D'Ovidio, R. Transgenic expression of polygalacturonase-inhibiting proteins in Arabidopsis and wheat increases resistance to the flower pathogen Fusarium graminearum. *Plant Biol.* **2012**, *14* (Suppl. 1), 31–38. [CrossRef]
82. Grant, G.T.; Morris, E.R.; Rees, D.A.; Smith, P.J.C.; Thom, D. Biological interactions between polysaccharides and divalent cations: The egg-box model. *FEBS Lett.* **1973**, *32*, 195–198. [CrossRef]
83. Limberg, G.; Körner, R.; Buchholt, H.C.; Christensen, T.M.; Roepstorff, P.; Mikkelsen, J.D. Quantification of the amount of galacturonic acid residues in blocksequences in pectin homogalacturonan by enzymatic fingerprinting with exo- and endo-polygalacturonase II from Aspergillus niger. *Carbohydr. Res.* **2000**, *327*, 321–332. [CrossRef]
84. Wu, H.-C.; Bulgakov, V.P.; Jinn, T.-L. Pectin Methylesterases: Cell Wall Remodeling Proteins Are Required for Plant Response to Heat Stress. *Front. Plant Sci.* **2018**, *9*, 1612. [CrossRef] [PubMed]
85. Wang, L.; Hart, B.E.; Khan, G.A.; Cruz, E.R.; Persson, S.; Wallace, I.S. Associations between phytohormones and cellulose biosynthesis in land plants. *Ann. Bot.* **2020**, *126*, 807–824. [CrossRef] [PubMed]
86. Díaz-Valle, A.; López-Calleja, A.C.; Alvarez-Venegas, R. Enhancement of Pathogen Resistance in Common Bean Plants by Inoculation With Rhizobium etli. *Front. Plant Sci.* **2019**, *10*, 1317. [CrossRef] [PubMed]
87. Vincken, J.P.; Schols, H.A.; Oomen, R.J.F.J.; McCann, M.C.; Ulvskov, P.; Voragen, A.G.J.; Visser, R.G.F. If homogalacturonan were a side chain of rhamnogalacturonan I. Implications for cell wall architecture. *Plant Physiol.* **2003**, *132*, 1781–1789. [CrossRef]
88. Tang, W.; Li, A.-Q.; Yin, J.-Y.; Nie, S.-P. Structural characteristics of three pectins isolated from white kidney bean. *Int. J. Biol. Macromol.* **2021**. [CrossRef] [PubMed]
89. García-Angulo, P.; Willats, W.G.T.; Encina, A.E.; Alonso-Simón, A.; Álvarez, J.M.; Acebes, J.L. Immunocytochemical characterization of the cell walls of bean cell suspensions during habituation and dehabituation to dichlobenil. *Physiol. Plant.* **2006**, *127*, 87–99. [CrossRef]
90. Encina, A.; Sevillano, J.M.; Acebes, J.L.; Alvarez, J. Cell wall modifications of bean (*Phaseolus vulgaris*) cell suspensions during habituation and dehabituation to dichlobenil. *Physiol. Plant.* **2002**, *114*, 182–191. [CrossRef]
91. Cosgrove, D.J. Plant cell wall extensibility: Connecting plant cell growth with cell wall structure, mechanics, and the action of wall-modifying enzymes. *J. Exp. Bot.* **2016**, *67*, 463–476. [CrossRef]
92. Deepak, S.; Shailasree, S.; Kini, R.K.; Muck, A.; Mithöfer, A.; Shetty, S.H. Hydroxyproline-rich Glycoproteins and Plant Defence. *J. Phytopathol.* **2010**, *158*, 585–593. [CrossRef]

93. Mnich, E.; Bjarnholt, N.; Eudes, A.; Harholt, J.; Holland, C.; Jørgensen, B.; Larsen, F.H.; Liu, M.; Manat, R.; Meyer, A.S.; et al. Phenolic cross-links: Building and de-constructing the plant cell wall. *Nat. Prod. Rep.* **2020**, *37*, 919–961. [CrossRef]
94. Wei, G.; Shirsat, A.H. Extensin over-expression in Arabidopsis limits pathogen invasiveness. *Mol. Plant Pathol.* **2006**, *7*, 579–592. [CrossRef]
95. Tiré, C.; De Rycke, R.; De Loose, M.; Inzé, D.; Van Montagu, M.; Engler, G. Extensin gene expression is induced by mechanical stimuli leading to local cell wall strengthening in Nicotiana plumbaginifolia. *Planta* **1994**, *195*, 175–181. [CrossRef]
96. Merkouropoulos, G.; Shirsat, A.H. The unusual Arabidopsis extensin gene atExt1 is expressed throughout plant development and is induced by a variety of biotic and abiotic stresses. *Planta* **2003**, *217*, 356–366. [CrossRef] [PubMed]
97. Merkouropoulos, G.; Barnett, D.C.; Shirsat, A.H. The Arabidopsis extensin gene is developmentally regulated, is induced by wounding, methyl jasmonate, abscisic and salicylic acid, and codes for a protein with unusual motifs. *Planta* **1999**, *208*, 212–219. [CrossRef]
98. Hückelhoven, R. Cell Wall–Associated Mechanisms of Disease Resistance and Susceptibility. *Annu. Rev. Phytopathol.* **2007**, *45*, 101–127. [CrossRef] [PubMed]
99. Betsuyaku, S.; Katou, S.; Takebayashi, Y.; Sakakibara, H.; Nomura, N.; Fukuda, H. Salicylic Acid and Jasmonic Acid Pathways are Activated in Spatially Different Domains around the Infection Site during Effector-Triggered Immunity in Arabidopsis thaliana. *Plant Cell Physiol.* **2018**, *59*, 8–16. [CrossRef]
100. Escudero, V.; Torres, M.Á.; Delgado, M.; Sopeña-Torres, S.; Swami, S.; Morales, J.; Muñoz-Barrios, A.; Mélida, H.; Jones, A.M.; Jordá, L.; et al. Mitogen-activated protein kinase phosphatase 1 (MKP1) negatively regulates the production of reactive oxygen species during Arabidopsis immune responses. *Mol. Plant-Microbe Interact.* **2019**, *32*, 464–478. [CrossRef] [PubMed]
101. R Development Core Team. *R: A Language and Environment for Statistical Computing;* R Foundation for Statistical Computing: Vienna, Austria, 2020; ISBN 3900051070.
102. Lê, S.; Josse, J.; Husson, F. FactoMineR: An R package for multivariate analysis. *J. Stat. Softw.* **2008**. [CrossRef]
103. Updegraff, D.M. Semimicro determination of cellulose inbiological materials. *Anal. Biochem.* **1969**, *32*, 420–424. [CrossRef]
104. Saeman, J.F.; Moore, W.E.; Millet, M.A. Sugar units present. Hydrolysis and quantitative paper chromatography. *Methods Carbohydr. Chem.* **1963**, *3*, 54–69.
105. Dubois, M.; Gilles, K.A.; Hamilton, J.K.; Rebers, P.A.; Smith, F. Colorimetric Method for Determination of Sugars and Related Substances. *Anal. Chem.* **1956**, *28*, 350–356. [CrossRef]
106. Blumenkrantz, N.; Asboe-Hansen, G. New method for quantitative determination of uronic acids. *Anal. Chem.* **1973**, *5*, 484–489. [CrossRef]
107. Albersheim, P.; Nevins, D.J.; English, P.D.; Karr, A. A method for the analysis of sugars in plant cell-wall polysaccharides by gas-liquid chromatography. *Carbohydr. Res.* **1967**, *5*, 340–345. [CrossRef]
108. De Castro, M.; Largo-Gosens, A.; Alvarez, J.M.; García-Angulo, P.; Acebes, J.L. Early cell-wall modifications of maize cell cultures during habituation to dichlobenil. *J. Plant Physiol.* **2014**, *171*, 127–135. [CrossRef] [PubMed]
109. Schindelin, J.; Arganda-Carreras, I.; Frise, E.; Kaynig, V.; Longair, M.; Pietzsch, T.; Preibisch, S.; Rueden, C.; Saalfeld, S.; Schmid, B.; et al. Fiji: An open-source platform for biological-image analysis. *Nat. Methods* **2012**, *9*, 676–682. [CrossRef]

*plants*

MDPI

*Review*

# Emerging Roles of β-Glucanases in Plant Development and Adaptative Responses

**Thomas Perrot, Markus Pauly and Vicente Ramírez ***

Institute for Plant Cell Biology and Biotechnology—Cluster of Excellence on Plant Sciences, Heinrich Heine University Düsseldorf, 40225 Düsseldorf, Germany; thomas.perrot@hhu.de (T.P.); m.pauly@hhu.de (M.P.)
* Correspondence: ramirezg@uni-duesseldorf.de

**Abstract:** Plant β-glucanases are enzymes involved in the synthesis, remodelling and turnover of cell wall components during multiple physiological processes. Based on the type of the glycoside bond they cleave, plant β-glucanases have been grouped into three categories: (i) β-1,4-glucanases degrade cellulose and other polysaccharides containing 1,4-glycosidic bonds to remodel and disassemble the wall during cell growth. (ii) β-1,3-glucanases are responsible for the mobilization of callose, governing the symplastic trafficking through plasmodesmata. (iii) β-1,3-1,4-glucanases degrade mixed linkage glucan, a transient wall polysaccharide found in cereals, which is broken down to obtain energy during rapid seedling growth. In addition to their roles in the turnover of self-glucan structures, plant β-glucanases are crucial in regulating the outcome in symbiotic and hostile plant–microbe interactions by degrading non-self glucan structures. Plants use these enzymes to hydrolyse β-glucans found in the walls of microbes, not only by contributing to a local antimicrobial defence barrier, but also by generating signalling glucans triggering the activation of global responses. As a counterpart, microbes developed strategies to hijack plant β-glucanases to their advantage to successfully colonize plant tissues. This review outlines our current understanding on plant β-glucanases, with a particular focus on the latest advances on their roles in adaptative responses.

**Keywords:** β-glucanases; β-glucans; cell wall polysaccharides; plant development; environmental stress

**Citation:** Perrot, T.; Pauly, M.; Ramírez, V. Emerging Roles of β-Glucanases in Plant Development and Adaptative Responses. *Plants* **2022**, *11*, 1119. https://doi.org/10.3390/plants11091119

Academic Editors: Penélope García-Angulo and Asier Largo-Gosens

Received: 28 March 2022
Accepted: 18 April 2022
Published: 20 April 2022

**Publisher's Note:** MDPI stays neutral with regard to jurisdictional claims in published maps and institutional affiliations.

**Copyright:** © 2022 by the authors. Licensee MDPI, Basel, Switzerland. This article is an open access article distributed under the terms and conditions of the Creative Commons Attribution (CC BY) license (https://creativecommons.org/licenses/by/4.0/).

## 1. Introduction

The plant cell wall is a dynamic composite structure with diverse functions including mechanical and structural support for plant growth, compartmentalization of specialized cells, and the integration of cell-to-cell communication and interactions with the environment [1,2]. Given these essential roles, plants have evolved intricate mechanisms to assemble, modify, and dismantle the diverse wall components. β-glucans including cellulose, callose, xyloglucan, and mixed-linkage glucan (MLG) are prevalent wall structural constituents in most plant cell types. These polysaccharides share a backbone composed of D-glucopyranosyl building blocks linked by β-1,4 and/or β-1,3 bonds, which can contain additional sidechain substitutions. The abundance, length, and associations of these glucan structures are modified during development and in response to environmental cues [3,4]. The diversification of glucan structures found in plant walls has been accompanied by the coevolution of specific hydrolases allowing for their modification/degradation. Plant genomes encode diverse types of enzymes able to hydrolyse β-glucans. Endo-β-glucanases, the most abundant type in plants, randomly cleave internal β-D-glucosidic linkages in the glucan backbone, while exo-β-glucanases act processively on both ends of the glucan chain, releasing oligosaccharides. Finally, β-glucosidases hydrolyse terminal β-D-glucosyl residues releasing β-D-glucose. These enzymes can be further classified in many ways depending on the reaction mechanism used, the chemical reaction they catalyse, or amino acid sequence-related aspects. Based on common amino acid sequences and protein fold structures, β-glucanases are typically distributed in multiple glycoside hydrolase (GH) families according to the classification of Carbohydrate-Active enZymes (CAZy) [5]. Frequently, Enzyme Commission (EC) number(s) are also used to designate the biochemical reaction(s) catalysed by these proteins [6]. Depending on the glycoside bond they hydrolyse, three main types of β-glucanases can be found in plants: β-1,4-glucanases, β-1,3-glucanases and β-1,3-1,4-glucanases.

1,4-β-glucanases are enzymes able to hydrolyse the 1,4-glycosidic bond between two contiguous D-glucopyranose units (Figure 1). This bond is found in the structure of cellulose, the most abundant polysaccharide in plant walls. 1,4-β-glucanases, initially termed "cellulases", include several types of enzyme activities needed to degrade cellulose, such as endoglucanases (EC 3.2.1.4), cellobiohydrolases (EC 3.2.1.91), and glucosidases (EC 3.2.1.21) [7–9]. Plant cellulases are classified in the GH9 family [5,10]. Although most of them exhibit only limited activity on crystalline cellulose, they are able to hydrolyse amorphous regions of cellulose and soluble cellulose derivatives such as carboxymethyl cellulose. There is some evidence suggesting that these enzymes are also able to cleave other non-cellulosic polysaccharides containing contiguous (1,4)-β-glucosyl residues in their backbone, including MLG, xyloglucan, and glucomannan, although characterization of the hydrolytic activities on different substrates has been limited to a handful of plant GH9 proteins [11]. GH9 β-1,4-glucanases have been implicated in several aspects of cell wall metabolism in higher plants, including cellulose biosynthesis and degradation, modification of the association of cellulose microfibrils with other wall polysaccharides, or wall loosening during cell elongation [12–14].

The term β-1,4-glucanase includes additional enzyme activities such as those involved in the hydrolysis of xyloglucans, heavily xylose-substituted 1,4-β-glucans [15]. Xyloglucan endohydrolases (XEH; EC 3.2.1.151) cleave the xyloglucan chain, releasing oligosaccharides. However, some of these enzymes also exhibit transglucosylase activity (XET; EC 2.4.1.207) being able to covalently link these oligosaccharides onto the non-reducing terminal end of the glucose moiety of other xyloglucan and cellulose acceptors [16–19]. This group of enzymes, generally termed xyloglucan endotransglucosylase/hydrolases (XTH) has important roles in wall polymer rearrangement, polymer integration into the wall, and cell expansion, and has also been implicated in plant responses towards abiotic and biotic stresses. Several reviews have recently addressed this transglycosylase class of plant β-1,4-glucanases [20,21], and they will not be further considered in this review.

β-1,3-glucanases catalyse the hydrolysis of glucans containing contiguous β-1,3-linked glucosyl residues (Figure 1; [22]). β-1,3-glucanases are responsible for the degradation of callose in plants, but they can also hydrolyse the β-1,3- and β-1,3-1,6-glucans found in the walls of intruding fungi [23,24]. They are also called laminarinases, as they are able to cleave laminarin, a linear β-1,3-glucan displaying occasional β-1,6-branches found in brown algae [25]. Laminarin and the structurally similar β-1,3-glucans paramylon and pachyman are usually employed to characterize in vitro activities of those enzymes. Plant β-1,3-glucanases are classified together with β-1,3-1,4-glucanases in the GH17 family of glycosyl hydrolases [26].

β-1,3-1,4-glucanases only hydrolyse β-1,4-glucosidic linkages if an adjacent β-1,3-glucosyl linkage is present towards the non-reducing end of the substrate (Figure 1). These enzymes are highly specific, and they are not able to hydrolyse β-1,3- or β-1,4-glucans [8,27,28]. In plants, these enzymes are involved in the degradation of MLG, a hemicellulosic polysaccharide prevalent in the walls of grass species. β-1,3-1,4-glucanases are also known as lichenases or licheninases, named after their activity on lichenin (moss starch), a complex glucan occurring in Parmeliaceae lichens [29,30]. This β-1,3-1,4-glucan differs from those characterized in grasses in having a much higher proportion of cellotriosyl to cellotetraosyl units [31]. In addition, despite having similar substrate specificities, plant β-1,3-1,4-glucanases have quite distinct amino acid sequences and 3D structures compared with microbial licheninases; thus, the term β-1,3-1,4-glucanase will be used [8,32,33].

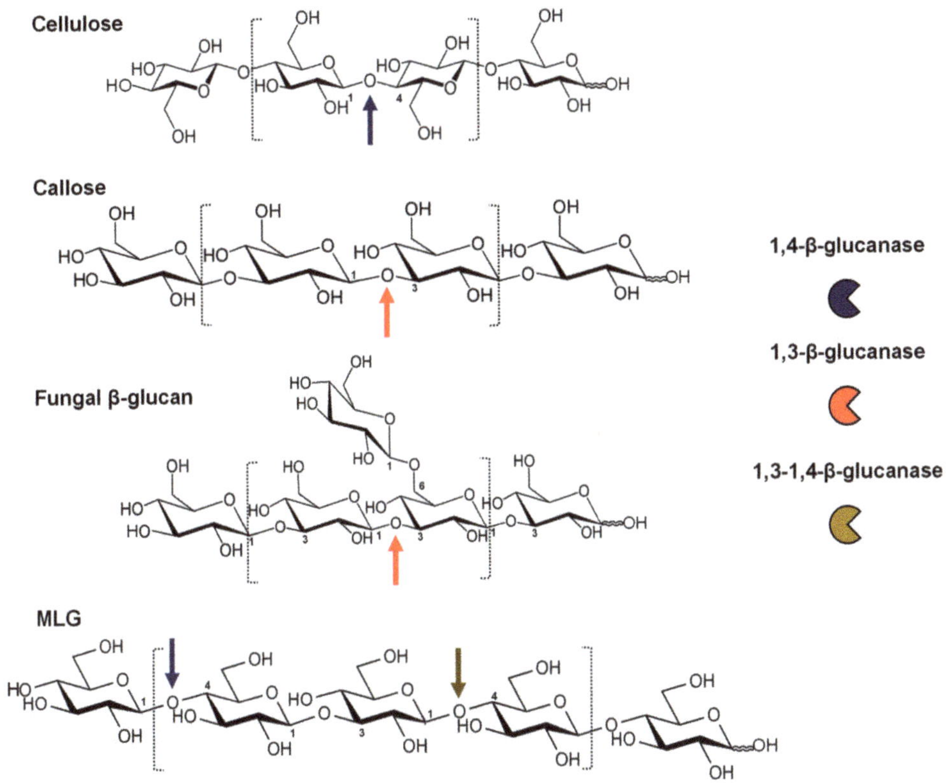


**Figure 1.** Activity of the three types of β-glucanases found in plants and their main physiological substrates. (1,3)- and (1,4)-β-glycosidic linkages are depicted as 3 and 4, respectively. Arrows indicate the glycosidic linkage hydrolysed in each case.

This review will encompass the evolutionary relationships, classification, and biological roles proposed for the following three types of plant endo-β-glucanases (referred as β-glucanases from here): β-1,4-glucanase ("cellulase" EC 3.2.1.4), β-1,3-glucanase (EC 3.2.1.39) and β-1,3-1,4-glucanase (EC 3.2.1.73). The specificities of these enzymes on their biological β-glucan substrates are shown in Figure 1.

## 2. β-1,4-Glucanases

### *2.1. Classification and Evolutionary Origin*

Enzymes able to use cellulose as substrate are found in many prokaryotic and eukaryotic organisms, including bacteria, fungi, nematodes, animals and plants [34]. However, the hydrolytic mechanisms used and thus the protein domains involved are different. In plants, a similar number of GH9 β-1,4-glucanases (EC 3.2.1.4) are found in diverse species such as Arabidopsis (25), poplar (25) barley (22), rice (24) sorghum (23) or Brachypodium (23) [34–37]. All these plant β-1,4-glucanases harbour a catalytic domain containing two conserved motifs required for the activity distinguishing GH9 from other hydrolases [9,38]. In the first motif, a histidine improves the stability of the substrate in the active site, while in the second motif, an aspartate (the nucleophile base) and a glutamate (the proton donor) are required for the cleavage of the glycosidic bond [39]. The presence of additional domains allows one to classify plant GH9 β-1,4-glucanases into three distinct structural subclasses: GH9A, GH9B and GH9C (Figure 2; [30–32]). In general, the domain structures and genomic intron–exon organization seem to be conserved in GH9 subgroups from distantly related plants such as monocots and dicots [36]. In the GH9A subgroup, the catalytic domain is preceded by a cytosolic N-terminal extension followed by a membrane-spanning domain that anchors the protein to the plasma membrane and/or to intracellular organelles. In subgroups GH9B and GH9C, the catalytic domain is instead preceded by a signal peptide for ER targeting and secretion. Subgroup GH9C members possess an additional long carboxyl-terminal extension containing a carbohydrate binding module (CBM49), which can bind cellulose in vitro (Figure 2; [36]).

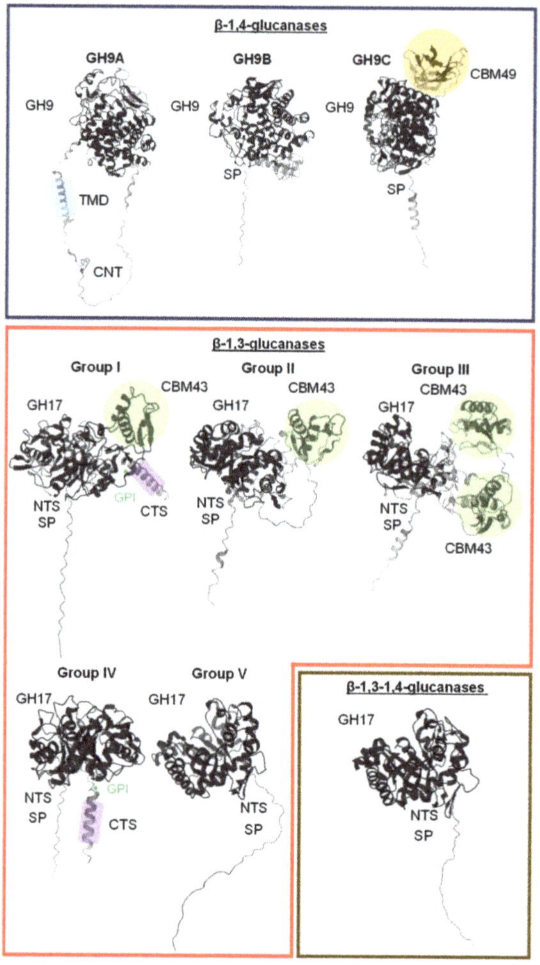


**Figure 2.** Topology models of plant β-glucanases. Protein domains and regions are abbreviated as follows: NTS—N-terminal sequence; GH—core glycosyl hydrolase family domain; CBM—carbohydrate binding module; GPI—glycosylphosphatidylinisotol-anchor attachment; CTS—hydrophobic C-terminal sequence; CNT—cytosolic N-terminal extension; SP—secretory signal peptide; TMD—transmembrane domain. Structure models represented in this figure were built by using the AlphaFold Protein Structure Database [40,41] with the following Uniprot accessions: GH9A 1,4-β-glucanase (AtGH9A1; Q38890), GH9B 1,4-β-glucanase (AtGH9B16; Q9SVJ4) and GH9C 1,4-β-glucanase (AtGH9C1; Q9M995), group I 1,3-β-glucanase (At5g64790; Q9LV98), group II 1,3-β-glucanase (At2g05790; F4IHD3), group III 1,3-β-glucanase (At2g39640; O48812), group IV 1,3-β-glucanase (At1g77780; Q9CA16), group V 1,3-β-glucanase (At5g20340; O49353), 1,3;1,4-β-glucanase (ZmMLGH1; B6T391).

Structural modelling and phylogenetic studies comparing the GH9 family in plant and non-plant taxa suggest that classes A and B of modern vascular land plants may have emerged by diverging directly from a common GH9C ancestor [42]. Although evidence is mostly based on in vitro activities and 3D modelling and should thus be interpreted with caution, some relationships between the three domain structures and biological functions have been proposed. The CBM49 domain, exclusive of plant GH9C β-1,4-glucanases, would confer the ability to digest crystalline substrates in a manner reminiscent from ancestral non-plant taxa such as bacteria. Subsequent loss of CBM49 during plant evolution and replacement of the signal peptide by a transmembrane region would have resulted in the appearance of the B and A classes, respectively. These events would have resulted in the development of new cellulose editing/modifying functions, where A and B classes would only be suited to digest amorphous regions of cellulose [42,43].

### 2.2. Biological Roles

Membrane-bound GH9A β-1,4-glucanases, lacking a CBM49 domain, would participate primarily in the assembly and repair/editing of cellulose microfibrils during cellulose biosynthesis in plants. Accordingly, KORRIGAN (KOR) GH9A β-1,4-glucanases are anchored to the plasma membrane associated with the cellulose synthase complex (CSC) and are required for the correct cellulose biosynthesis in both primary and secondary walls [14]. Several KOR genes have been identified in diverse plant species, KOR1 being the best studied member of this group [12,44]. In vitro, KOR1 is able to hydrolyse amorphous but not crystalline cellulose [45,46]. KOR1 β-1,4-glucanase activity is required for the efficient function of the CSC, and kor1 mutations result in CSC intracellular trafficking defects, reduced cellulose polymerization velocity, and aberrant crystalline cellulose deposition [13,14,47–50]. The precise mechanism is unclear, but it has been hypothesized that KOR1 would hydrolyse disordered amorphous glucan chains during cellulose synthesis. This "proofreading" activity might prevent unwanted interactions of microfibrils from adjacent CSCs, ensuring the correct fibril assembly. A second hypothesis is that KOR1 might cleave cellulose chains to terminate their synthesis once an adequate size is reached and before the CSC is internalized from the plasma membrane for recycling through endocytosis. A third possibility states that KOR1 is involved in the synthesis of a priming molecule required for cellulose synthesis initiation [45,51,52]. KOR1 function in cellulose biosynthesis seems to be conserved in several plant species including rice (*Oryza sativa*), tomato (*Solanum lycopersicon*), or *Populus* sp. [53–56]. Other GH9A genes with different expression profiles are also involved in cellulose synthesis in specific tissues and developmental conditions. For example, *Populus* PtrCel9A6 and PtrGH9A7 are two membrane-anchored β-1,4-glucanases which seem to regulate cellulose synthesis during xylem differentiation and lateral root formation, respectively [57,58].

Some defence-related defects have been associated with GH9A misregulation, although given the importance of a correct cellulose deposition in plants, these phenotypes are likely an indirect consequence of secondary metabolic alterations [55,59]. For example, the kor1-1 Arabidopsis mutant exhibits increased disease resistance to *Pseudomonas syringae* associated with an enhanced production of jasmonic acid, antagonizing the activation of salicylic acid-mediated defences [59]. Such an altered defence status has been further linked to an enhanced mycorrhization in KOR1-downregulated Populus deltoides plants [55].

GH9B β-1,4-glucanases are secreted into the apoplast where they have been typically associated with the wall remodelling in muro. Due to the lack of CMB49, these enzymes have limited or no hydrolytic activity on crystalline cellulose (Figure 2). Instead, GH9B could hydrolyse amorphous regions of cellulose involved in the interaction with other wall polymers altering the mechanical properties and promoting wall loosening or disassembly required in physiological processes such as cell expansion, fruit maturation or leaf abscission [9,60–65]. Evidence of biochemical activity for this type of enzyme is scarce, and their biological substrate(s) are usually inferred from indirect evidence. For example, a fruit-specific expression pattern has been observed for several β-1,4-glucanases in tomato, strawberry, avocado or pepper [60,66–68]. Fruits entering the ripening phase stop growing and tissues start softening, which is linked to wall structural alterations. FaEG1, a secreted GH9B β-1,4-glucanase, is specifically induced during strawberry ripening (Figure 2; [69]). Based on compatible cellulose- and xyloglucan-FaEG1 molecular docking simulations, it was suggested that FaEG1 might function in the disassembly of the cellulose–hemicellulose fraction during the ripening of strawberry fruit [69].

In tomato, silencing of the SlGH9B2 1,4-glucanase results in an increase in the force required to detach fruits from the plant [61]. Together with the localized expression in the abscission zone, it was suggested that SlGH9B2 is involved in wall disassembly at specific sites of organ detachment. Similarly, Arabidopsis Cel3 and Cel5 GH9Bs are thought to promote wall loosening required for the detachment of root cap cells during root development and in response to biotic and abiotic rhizosphere stresses [63–65].

In poplar, it was hypothesized that Cel1 and Cel2 promote leaf growth by hydrolysing non-crystalline regions of cellulose and by releasing xyloglucans intercalated within the disordered domains of cellulose microfibrils. This would reduce wall stress, increasing the accessibility of wall components to additional enzymatic modifications required for wall restructuring during cell expansion and growth [62,70–72]. Accordingly, overexpression of the poplar Cel1 1,4-β-glucanase decreases the amount of xyloglucan cross-linked with cellulose microfibrils, increasing wall plasticity and promoting enlargement of plant cells [72]. Similarly, overexpression of OsGH9B1 and OsGH9B3 1,4-β-glucanases in rice results in reduced cellulose crystallinity and degree of polymerization. Although only slight differences in wall polymer composition, stem mechanical strength, and biomass yield were observed, GH9B1/3 overexpression enhanced wall enzymatic digestibility likely caused by an enhanced accessibility due to a more open wall architecture [73].

GH9B 1,4-β-glucanases have also been involved in the maintenance of secondary wall mechanical strength [74]. GH9B5 is highly expressed during secondary wall development in poplar and Arabidopsis. AtGH9B5 t-DNA knock-out mutant walls exhibit decreased mechanical strength accompanied by reduced xylose and glucose content in the hemicellulosic fraction, while overexpression of AtGH9B5 using a promoter specific to secondary wall development led to the opposite phenotype with an increase in cellulose crystallinity. According to these results, it was proposed that GH9B5 might facilitate the coating of cellulose microfibrils with xylans during secondary wall formation resulting in strengthened walls [74]. Unlike other GH9B members, GH9B5 contains a predicted GPI anchor, thus linking the GH initially to the plasma membrane [75]. Proteomic analyses are consistent with such an association with the plasma membrane, potentially bringing GH9B5 in close contact with the CSC (Figure 2; [74,76,77]).

A new role for a selected clade of secreted GH9B β-1,4-glucanases has recently been proposed in the cell-to-cell adhesion process during grafting and parasitic plant–host interaction. Successful grafting requires the adhesion of facing cells at the graft boundary [78]. NbGH9B3 is induced together with other wall remodelling genes in these graft junctions. NbGH9B3 misregulation results in a significant decrease in the graft success percentage in multiple inter- and intra-specific grafts [79]. Although the exact substrate involved is still unknown, it has been proposed that NbGH9B3 might target amorphous cellulose regions at the graft boundary, promoting cell adhesion and the formation of xylem bridges required for a compatible graft [79]. The expression of similar GH9B β-1,4-glucanases during compatible grafting is conserved across other plant species such as soybean (*Glycine max*), morning glory (*Ipomoea nil*), or Arabidopsis, suggesting a common mechanism.

Some parasitic plants seem to use a comparable mechanism to disrupt host cell wall barriers and develop similar xylem bridges, allowing the parasitic plant to colonize host tissues. During *Phtheirospermum japonicum–Arabidopsis* interaction, *P. japonicum* secretes the PjGH9B3 ortholog in the periphery of the haustorium in direct physical contact with the host tissue. As a result, the thickness of the host wall at the interface area between both parasite and host plant cells is decreased, and xylem bridges are formed, allowing a successful cell-to-cell adhesion and further colonization [80].

Alterations in the wall structure caused by GH9Bs have also been involved in the interaction between plants and microbial pathogens. Arabidopsis GH9B1 and GH9B2 and their respective tomato homologs are required for the establishment of balanced defence responses against pathogens with different lifestyles. Although the exact role is unclear, it has been speculated that a lack of SlGH9B1/2 might modify the structure and properties of the wall, altering its signalling abilities [81,82].

Expression of specific GH9B β-glucanases has also been associated with a more efficient defensive response against nematode attacks. According to the proposed mechanism, plants activate the expression of secreted β-glucanases at the nematode feeding site to halt the infection, likely interfering with pathogen development [83]. For example, AtGH9B Cel6 is highly expressed during compatible plant–nematode interactions, and ectopic expression of AtCel6 in soybean roots reduces disease susceptibility to soybean cyst nematode (*Heterodera glycines*) and root knot nematode (*Meloidogyne incognita*) [83].

Some nematodes have acquired the ability to recruit plant β-glucanases to the infection sites as a way to promote virulence by assisting in the development of giant cells and syncytia, specialized feeding structures [83,84]. Although the formation of these structures is different depending on the species, it often involves extensive remodelling of both host and nematode walls, allowing for cell elongation and the appearance of elaborate ingrowths aiming to increase the surface area for nutrient uptake [84–86]. Evidence suggests that most of these wall modifications arise from the action of plant enzymes (e.g., β-glucanases), rather than enzymes of nematode origin. Expression of GmCel7, a homolog of Arabidopsis Cel2 (GH9B subclass; Figure 2), is highly induced in nematode-infected roots. While suppression of GmCel7 did not seem to modify the plant wall architecture under normal developmental conditions, it reduced the amount of feeding structures and mature *H. glycines* females present in infected roots by 50% [83]. Although its substrate specificity is unknown, these results suggest that GmCel7 is required for the full virulence of *H. glycines*. The specific induction GH9B and GH9C plant β-glucanases during plant–nematode interactions is well documented in soybean, tobacco and Arabidopsis, suggesting that the mechanism might be an extended nematode virulence strategy [83,84,87–89].

CBM49-containing β-1,4-glucanases (GH9C) are thought to be involved in the degradation of crystalline cellulose associated with irreversible wall disassembly such as root hair emergence or breakdown of the endosperm wall during germination [90–92]. Experimental evidence shows that the CBM49 domain from the tomato β-1,4-glucanase Cel9C binds crystalline cellulose substrates in vitro [92]. Upregulation of the Cel9C transcript has been observed in ripening fruits correlated with irreversible wall degradation. Although the exact physiological role of SlGH9C1 remains elusive, it could function complimentary with GH9B β-1,4-glucanases and other wall-degrading enzymes during the fruit ripening phase.

Structural studies on characterised and putative GH9C class enzymes suggest that CBM49 might be contributing to substrate selection/modification rather than direct hydrolysis of crystalline cellulose [43]. Site-directed mutagenesis highlighted the importance of some aromatic residues in CBM49, contributing to the interaction with the surface of crystalline substrates [92]. Similar residues in homologous CBMs from non-plant GH9 β-1,4-glucanases are important to discriminate between related ligands such as cellulose, xylan, or MLG [93–97]. Although cellulose is the likely physiological substrate for GH9C plant 1,4-glucanases, it cannot be excluded that they are also able to degrade other non-cellulosic polysaccharides in muro [43,91,92]. Supporting this idea, the catalytic domains of SlCel9C1 and its rice ortholog show broad substrate specificity and are able to hydrolyse not only crystalline cellulose but also arabinoxylan or MLG [92,98].

In Arabidopsis, AtGH9C1 knock-down mutants show defects in root hair formation and delayed germination. AtGH9C1 is primarily expressed in root hair cells and endosperm cells, two cell types with specialized walls which are fully dismantled during development [91]. It is likely that AtGH9C1 hydrolyses cellulose in these wall types to facilitate wall weakening, although enzyme activity or substrate preference have not been reported. The presence of the CBM49 domain seems to target AtGH9C1 to particular regions of the wall, and when it is missing the protein, it is diffusely distributed throughout the wall [91]. In a different study, it was observed that the CBM domain of a rice GH9C endoglucanase is post-translationally cleaved after the protein was targeted to the wall [98]. The proteolytic cleavage of CBM-containing GH9C proteins might activate their hydrolytic activity or modify their substrate specificity. After activation, GH9Cs could regulate the degree of crystallinity and thus the cross-linking with matrix polysaccharides, controlling cell expansion during development [74]. Downregulation of AtGH9C2 led to decreased cellulose crystallinity associated with an increase in plant height and rosette diameter. Conversely, overexpression of the poplar ortholog PtGH9C2 results in increased cellulose crystallinity and opposite plant phenotypes [74]. Considering the characteristic expression profile of GH9C2 in Arabidopsis and Poplar, it has been proposed that this β-1,4-glucanase controls cell expansion during secondary wall development by regulating the crystallization of cellulose microfibrils and thereby the interaction with other wall polymers [74].

## 3. β-1,3-Glucanases

### 3.1. Classification and Evolutionary Origin

The large diversity of β-1,3-glucanases makes their classification difficult, resulting in a widespread distribution in several clades of the glycoside hydrolase superfamily [99]. Several attempts have been made to classify β-1,3-glucanases in different plant species. Initial classifications were based on sequence similarity, molecular size, or isoelectric point. As a result, a number of distinct β-1,3-glucanases classes have been defined in tobacco [100], barley [8], cotton [101], soybean [102] or rice [103,104]. Despite sharing a significant protein structural similarity and sequence-related attributes, members of the same subgroups showed a high diversity of temporal and tissue-specific expression patterns. For example, 27 rice β-1,3-glucanases were classified into four groups based on amino acid sequence [104]. Enzymes classified within the same group were differentially expressed in various tissues, developmental stages as well as in response to different plant biotic and abiotic stresses and hormone treatments. While OsGlu11 and 27 were specifically expressed in stem and leaf and strongly induced by *Magnaporthe grisea* infection and SA and ABA treatments, several other members of the same family were expressed in flowers and germinating grains [104]. Such classifications allowed for a good separation of β-1,3-glucanases from the sequence-related β-1,3;1,4-glucanases encoded in the same genome [99]. However, the diversity of expression patterns of members of the same group did not allow one to infer common biological roles.

In an attempt to clarify the biological significance of β-1,3-glucanase gene multiplicity and the relationship between sequence and function, more complex classification methods have been established [105,106]. Advanced phylogenetic analysis combined with expression information, knowledge of co-expressed genes, and published data allowed for the grouping of 50 putative β-1,3-glucanases encoded in the Arabidopsis genome in five protein domain architectures (Group I to V; Figure 2). The exon–intron organization of the corresponding genes within these five protein structural archetypes is relatively conserved in distantly related species such as cotton, cacao, and grape vine [106]. The common protein domains shared among these five groups consist of the presence of an N-terminal sequence (NTS) and a core glycosyl hydrolase (GH) domain. Group V is composed by proteins with this basic architecture. The remaining four groups contain additional domains. A carbohydrate-binding module (CBM43) and a hydrophobic C-terminal sequence (CTS) are present after the GH domain in members of group II and IV, respectively. The CBM43 domain is known to allow for the binding to β-1,3-glucans [107], while the CTS sequence encodes a transient transmembrane domain similar to vacuolar targeting peptides and glycosylphosphatidylinisotol (GPI)-anchor attachments [108–110]. Group I proteins contain both the CBM43 domain and the CTS region. Finally, group III proteins have a second CBM43 domain, but they do not contain the CTS region (Figure 2).

There seems to be a series of evolutionary events leading to the divergence in β-1,3-glucanase protein domain architectures and expression patterns to predict biological functions [105]. According to this model, ancestral β-1,3-glucanases belonging to the architecture group I show an abundant expression in a variety of tissues and organs and might have a predominant role in wall remodelling during plant development. Additional emerging β-1,3-glucanases (groups II to V) would have appeared during evolution by gene-duplication, subsequent loss of protein domains, and regulatory cis-elements leading to the observed structural and functional diversity. For example, stress-related β-1,3-glucanases would have evolved from group I ancestors with the acquisition of stress-responsive expression patterns followed by loss of the CTS region including the GPI-anchoring site, thus allowing for extracellular secretion upon certain signals. Accordingly, Arabidopsis group II (lacking the CTS domain) is enriched in pathogenesis-related β-1,3-glucanases, whose expression is specifically induced in response to pathogens and thus might be involved in the degradation of β-1,3 glucans present in fungal cell walls [105,106].

### 3.2. Biological Roles

Plant β-1,3-glucanases have received significant attention as important players in plant–microbe interactions, as a large number of β-1,3-glucanases are included in pathogenesis-related (PR) group 2 of proteins [24]. This type of β-1,3-glucanases accumulates in the event of a pathogen attack, and some of them have been directly involved in the hydrolysis of pathogen walls, as β-1,3-glucans are found in bacteria, metazoa, viruses, and particularly fungi. In this last group, β-1,3-glucans constitute the most abundant wall structural component existing in the form of linear or branched glucans containing mostly β-1,3 and β-1,6 linkages [111–113]. Antifungal hydrolytic activity of β-1,3-glucanases isolated from multiple plant species has been confirmed both in vitro and in vivo.

Purified PR-2 β-1,3-glucanases often show inhibitory effects on fungal spore germination and on hyphal growth, causing mycelial deformations and lysis due to the hydrolysis of the fungal walls. This type of antifungal activity was shown by the tobacco GluII β-1,3-glucanase against *Fusarium solani* [114]. The antifungal activity of β-1,3-glucanases is often not species specific, and it show inhibitory effects on a wide range of pathogens. For example, extracts from pea plants overexpressing a barley β-1,3-glucanase reduces the spore germination of *Trichoderma harzianum* and *Colletotrichum acutatum* and delays hyphal growth in *Botrytis cinerea* and *Ascochyta pisi* [115]. Similarly, a wheat β-1,3-glucanase heterologously produced shows inhibitory effects on hyphal growth, spore formation and mycelial morphology of *Fusarium*, *Alternaria* and *Penicillium* species [116]. Furthermore, some β-1,3-glucanases have shown antifungal activity against pathogens specialized in infecting distantly related plant clades. For example, wheat TaGluD showed in vitro antifungal activity not only against monocot-specific *Rhizoctonia pathogenic* strains, but also against *Phytophthora capsici* and *Alternaria longipes*—specialized pathogens infecting dicot hosts such as hot pepper and tobacco, respectively [117].

Direct antifungal effects have also been observed in vivo. For example, functional studies of rice–*Magnaporthe oryzae* interactions demonstrated a direct antifungal effect of the Gns6 β-1,3-glucanase dependent on its hydrolytic activity [118]. While wildtype Gns6 was able to inhibit conidial germination and formation of appressoria, a non-catalytic gns6 mutant lost all antifungal effects, showing enhanced pathogen colonization and increased disease susceptibility [118]. Furthermore, differential expression of PR-2 β-1,3-glucanases has been associated with increased disease resistance when comparing susceptible and resistant cultivars in a number of crop species such as rocket salad, tomato, and wheat [119–121]. Transgenic expression of β-1,3-glucanase genes has been extensively used as a strategy to develop durable disease resistance against fungal pathogens in crop plants [24]. For example, overexpressing a tobacco β-1,3-glucanase in transgenic groundnut plants significantly improved disease resistance against the pathogens *Cercospora arachidicola* or *Aspergillus flavus*, reducing the number and size of the necrotic lesions and a general delay in the disease symptom progression or preventing aflatoxin accumulation in seeds, respectively [122]. Similar protective effects have been reported by overexpressing potato, barley, or soybean β-1,3-glucanases in flax, wheat, and banana transgenic lines [123–125]. In general, these PR-2 β-1,3-glucanases belong to the Group V, characterized by the presence of a secretion signal and the absence of CBM and CTS domains (Figure 2).

Plant β-1,3-glucanases are also involved in an alternative plant defence mechanism, where degradation of the hyphal cell wall of invading fungi would release β-1,3-glucan oligosaccharides serving as elicitors. Upon recognition by plant surveillance systems, these β-1,3-glucan oligosaccharides would act as microbe-associated molecular patterns (MAMPs) triggering the activation of signalling cascades resulting in a wide range of localized and systemic defence responses [126]. For example, upon oomycete infection, soybean and rice β-1,3-glucanases are able to release β-glucan elicitors from the fungal wall, which trigger the production of antimicrobial compounds such as phytoalexins [127–129]. Likewise, a synthetic β-1,3/-1,6-glucan heptaglucoside similar to those obtained after β-1,3-glucanase hydrolysis from fungal walls is able to activate plant defences in legume species [130,131]. The exact mechanism of recognition and mode of action of pathogen-derived β-glucan elicitors is not fully understood, and it seems to be dependent not only on the plant species and the origin of the β-glucan elicitor, but also on the glycan structure, where length, branching pattern and presence of chemical modifications have shown to be relevant [132]. Recent evidence suggests that this mechanism might be more widespread among the plant kingdom than expected, and diverse β-1,3 glucanase-derived oligosaccharides were recently demonstrated to be potent elicitors of plant defences in *Nicotiana benthamiana*, barley, rice, and *Arabidopsis* [133,134].

In addition to the generation of β-glucan elicitors, recent evidence suggests that plant β-1,3-glucanases are also involved in the release of soluble β-glucans from fungal walls with antioxidant activities required not only for pathogenicity, but also for the successful establishment of root interactions with beneficial endophytic fungi [135]. Barley HvBGLUII, an apoplastic Group V β-1,3-glucanase, is able to partly hydrolyse Serendipita indica walls. As a result, a non-immunogenic β-1,3;1,6-glucan decasaccharide (β-GD) with ROS scavenging properties is released, thus facilitating fungal development. Exogenous applications of β-GD increase the fungal root colonization efficiency. In addition, these effects are highly specific to the length and branching pattern of the oligosaccharide, as altering the glucan structure blocks its activity completely. It has been proposed that such a hijacking of plant apoplastic β-1,3-glucanases might be a common fungal counter-defensive strategy to subvert host immunity during pathogen and beneficial fungi colonization [135]. Given the heterogeneity of 1,3- and 1,3-1,6-glucan structures found in fungi and other microbes, the number of biological processes that involve such glucan effectors resulting from the hydrolytic action of plant β-glucanases might be larger than previously considered.

In addition to their roles during plant–microbe interactions β-1,3-glucanases also play a role in the degradation of plant produced callose, a transient β-1,3-glucan polymer accumulated at discrete sites during certain developmental processes and in response to environmental cues. For example, during microsporogenesis, another specific β-1,3-glucanases degrades the thick callose wall surrounding the tetrad, releasing the microspores into the anther locule for pollen maturation [136–138]. Rice mutants harbouring a catalytically inactive Osg1 β-1,3-glucanase produce damaged pollen grains. In those mutant pollen grains, large amounts of callose remains, and microspores remain attached to each other resulting in deformations and male sterility [139]. In a less understood way, stigma- and style-specific extracellular β-1,3-glucanases have been involved in pollination, specifically in the degradation of callose in the stylar matrix during pollen tube growth [140].

Another example of the involvement of β-1,3-glucanases in plant development is the control of callose degradation in plasmodesmata (Pd). Pd form channels physically interconnecting the cytoplasm and endoplasmic reticulum of adjacent cells [141,142]. β-1,3-glucanase-mediated callose degradation is thought to be tightly regulated in order to control the symplastic trafficking of micro- and macromolecules, including phytohormones, mRNA, and mobile non-cell-autonomous transcriptional factors [143,144]. Controlled build-up and degradation of callose deposits in Pd is relevant for multiple developmental processes such as germination, axillary bud growth, lateral root organogenesis, shoot apical meristem determination, or stomatal differentiation [145–149]. In addition, degradation of callose in Pd has been observed in response to environmental signals such as low temperatures, circadian rhythm or pathogen attack [141]. This type of Pd-specific β-1,3-glucanase contains a GPI-anchor targeting them to the plasma membrane. Moreover, they can also have a CBM43 domain, but its presence is facultative. Hence, they are classified into groups I or IV (Figure 2). They show a characteristic punctuate pattern around Pd colocalizing with callose deposits. In Arabidopsis, Pd-localized β-1,3-glucanase knockout mutants contain larger callose deposits, causing a decrease in Pd conductivity and cell-to-cell transport defects [143,146]. In perennial plants such as birch (*Betula pubescens*) or tree peonies (*Paeonia suffruticosa*), breakage of bud dormancy after the winter season is mediated by a Pd-specific β-1,3-glucanase in charge of degrading callose deposits and thus restoring symplasmic organization [150,151]. Two CBM43-containing enzymes—PsBG6 and PsBG9—have been implicated in this process. Despite both of them having been Pd-localized, PsBG6 was found inside of the Pd passage, whereas PsBG9 was observed to enclose the Pd channel, correlated with the presence of a GPI-anchor only in PsBG6 (Group I) and not in PsBG9 (Group II) (Figure 2; [151]).

In tobacco, a similar β-1,3-glucanase, ßGluI, regulates the release of coat-imposed dormancy and germination by promoting the rupture of the testa [152]. In tomato and banana, the expression of GPI-anchored plasmodesmata-located β-1,3-glucanases has been linked to fruit formation and softening through the modulation of symplasmic unloading of sugars and signalling molecules into the ovaries and fruit tissues [153,154].

Pd cell-to-cell communication channels are also utilized by viral pathogens to move and infect their plant hosts. β-1,3-glucanase mutants affected in the callose degradation in Pd show increased viral particle movement and enhanced susceptibility [155–157]. Several virus, bacteria and filamentous pathogen effectors have been shown to target Pd conductivity [158]. Potato virus Y is able to induce plant Pd-localized β-1,3-glucanase activity as an infection strategy resulting in the degradation of callose in Pd increasing viral movement and spread [159]. This β-1,3-glucanase upregulation and callose degradation is not observed in resistant potato cultivars where the virus fails to spread [160].

Callose is also accumulated in the form of cell wall thickenings at the site of contact with pathogens such as viruses, fungi or bacteria, and in wounds caused by herbivory attack. As part of the complex transcriptional reprograming regulated by the absiscic acid-jasmonic acid hormonal crosstalk, specific β-1,3 glucanases are downregulated during incompatible plant virus interactions [161,162]. Similarly, β-1,3-glucanase downregulation is observed during foliar fungal infections, where haustorial feeding structures are rapidly surrounded by callose deposits [163]. However, biochemical characterization of the β-1,3-glucanases involved in this type of defence-related callose deposition is scarce.

In winter cereals, the expression of Group V β-1,3-glucanases with anti-freezing activity has been associated with increased chilling tolerance [164]. For example, in rye (*Secale cereale* L.) freezing temperatures induce the expression of ScGLU-2 and ScGLU-3 β-1,3-glucanases. In vitro, these β-1,3-glucanases partially retain hydrolytic activity at sub-zero temperatures and have the ability to limit the growth of ice crystals. Structural modelling identified ice-binding surfaces (IBS) specifically in these β-1,3-glucanases, geometrically complementary to the surface of ice crystals. Residues on the putative IBSs are charge conserved in tolerant varieties but not in similar β-1,3-glucanases from non-acclimated rye varieties [165]. Similarly, accumulation of the VcGNS1 β-1,3-glucanase was observed in the skin of harvested table grapes after a prolonged cold treatment. Recombinant VcGNS1 is stable at 0 °C and is able to delay the inactivation of some enzymes after repeated freeze–thawing cycles [166]. Although it is not clear whether glucan hydrolytic activity is required for these effects, expression of β-1,3-glucanases with in vitro cryoprotectant activity has been correlated with chilling tolerance in species such as tobacco, tomato, and spinach [166–168].

## 4. β-1,3-1,4-Glucanases

Phylogenetic evidence supports the hypothesis that plant β-1,3-1,4-glucanases evolved from an ancestral Group V β-1,3-glucanase from the widely distributed family GH17 (Figure 2). The accumulation of only a limited number of point mutations during evolution, causing minor differences in the amino acid sequence at the substrate-binding and catalytic sites, was sufficient to develop this unique substrate specificity [8,32,169]. Thus, predictions of β-1,3-1,4-glucanase or β-1,3-glucanase activities based only on amino acid sequence information are often unsuccessful, and biochemical characterizations are required.

A handful of β-1,3-1,4-glucanases from monocot species have been characterized [170–176]. Often, cereal species encode several β-1,3-1,4-glucanase isoforms with different temporal and tissue-specific expression patterns. In cereals with an MLG-rich endosperm such as barley or rice seed-specific β-1,3-1,4-glucanase, expression is detected rapidly after seed imbibition and reaches a maximum at early stages during germination. In most cases, a second isoform is expressed in the grain endosperm during germination, but it is also detected during seedling elongation and in vegetative tissues such as leaves and roots. For example, two barley β-1,3-1,4-glucanases EI and EII have been characterized, showing high sequence similarity. While expression of EII is restricted to the scutellum during early gemination stages, EI expression is also detected in adult roots and leaves [170,177–180]. Similarly, rice EGL1 and EGL2 isoforms show a different spatiotemporal pattern [174] EGL2 being seed-specific and EGL1 expressed also in vegetative tissues, reaching maximum values at full expansion and then decreasing upon leaf aging.

### *Biological Roles*

Since expression of seed-specific β-1,3-1,4-glucanases is restricted to embryonic tissue, their function is likely to loosen the walls during the course of endosperm mobilization. During germination, breaking down this physical barrier is required to allow for access of additional hydrolytic enzymes to substrates within the cell, such as proteins or starch [170,181].

The functional role of β-1,3-1,4-glucanases in developing leaves and roots is less clear, and several hypotheses have been proposed. The asymmetrical conformation of MLG due to the irregular distribution of β-1,3-linkages limits the capacity of the polymer to aggregate into fibrillar structures such as cellulose. Instead, MLG forms a gel-like matrix able to closely associate with cellulose and xylan to reinforce the wall and guide the orientation of cellulosic microfibrils during growth [182–184]. As the developmental pattern of characterized β-1,3-1,4-glucanases in barley, rice, or wheat seemed to indicate that the enzymes accumulate in vegetative organs during the growing phase, it was hypothesized that β-1,3-1,4-glucanases could play a role in loosening the wall by partially hydrolysing MLG, allowing for turgor pressure-driven cell expansion [169,185,186]. However, several pieces of evidence do not support this hypothesis. An increase in the expression of EI is observed after transferring barley seedlings from a normal light/dark photoperiod to continuous darkness. Despite a detection of concomitant decrease in MLG abundance, there was no measurable elongation of leaves, suggesting that wall loosening does not occur [187]. Additionally, the dark-induced accumulation of EI and MLG degradation in leaves is strongly inhibited by high glucose availability in the growth media, indicating that β-1,3-1,4-glucanase activation might be uncoupled from cell expansion. Instead, it has been postulated that MLG could be used by plants as a short-term glucose supply and thus metabolizable energy stored in the wall [187]. According to this second hypothesis during dark cycles, low glucose levels would reach a certain threshold, triggering the induction of β-1,3-1,4-glucanase activity. As a consequence, MLG would be partially depolymerized, and the resulting oligosaccharides could be easily converted into glucose by accessory proteins. The released glucose would be immediately available within the plant providing a flexible energy source under sugar-depleting conditions. This relatively simple catabolic process would represent an advantage in the utilization of MLG as a carbon source over starch and other wall polysaccharides [181]. However, the recent characterization of the first β-1,3-1,4-glucanase mutants in maize and brachypodium raised some reservations regarding this hypothesis and revealed new aspects of MLG degradation in cereals [175,176]. In maize, MLG degradation in vegetative tissues is dependent on a single β-1,3-1,4-glucanase. Mixed Linkage Glucan Hydrolase 1 (MLGH1) has a high sequence homology and expression pattern compared to barley EI or rice EGL1. Although the maize genome encodes several genes annotated as putative β-1,3-1,4-glucanases, the sole disruption of MLGH1 blocks all detectable dark-induced MLG degradation in mutant seedlings [175]. An extended time course analysis showed that, similar to starch accumulation of MLG in maize, seedlings follows a circadian rhythm with the maximum MLG abundance observed at the end of the day and constant decrease once the dark period starts. Transcript accumulation of MLGH1 correlates with the degradation of MLG during these cycles. While mutant plants are not able to degrade MLG and no cycling is observed, MLGH1 overexpression avoids MLG accumulation, and only trace amounts of MLG can be detected [175]. Similarly, the single disruption of the brachypodium homolog, LCH1, results in loss of all detectable dark-induced MLG degradation in vegetative tissues [176]. These results indicate that there is no or limited functional redundancy among β-1,3-1,4-glucanases, and the existence of different isoforms encoded in the genome of individual species might respond to the need of stress- or spatiotemporal-specific expression patterns. Despite the impediment in MLG degradation maize and brachypodium β-1,3-1,4-glucanase, mutants do not exhibit significant developmental or growth-related defects showing similar plant height, vegetative biomass, and grain yield as wild-type plants, suggesting that the hydrolysis of MLG as a source or energy during the night seems not to be essential under normal growth conditions [175,176]. A possible explanation is that under these conditions, starch may serve as the major energy storage polymer, thus allowing mutant plants to maintain their growth rate despite being unable to utilize MLG as a carbon source. However, no significant differences were observed in starch degradation in dark-treated maize mlgh1 seedlings compared to the wild-type, and only a slightly faster starch turnover was reported in adult brachypodium lch1 mutant plants. One possibility is that MLG turnover makes a difference only in specific cell types or under specific circumstances such as plants growing under biotic/abiotic stresses. Hence, these β-1,3-1,4-glucanase mutants are valuable tools to address the biological significance of MLG in cereals in the future.

The implication of diverse plant β-1,3-1,4-glucanases in stress responses such as plant–pathogen interactions has long been proposed, and new evidence has emerged. The expression of some β-1,3-1,4-glucanase genes in rice is induced by wounding, infection with virulent strains of the blast fungus (*Magnaporthe grisea*), and treatment with salicylic acid or methyl jasmonate defence-related phytohormones [103,174,188]. Not surprisingly, rice plants overexpressing the GNS1 β-1,3-1,4-glucanase show an increased disease resistance to *M. grisea*. These mutant plants form spontaneous brown specks accompanied by constitutive activation of defence-related genes and stunted growth [174]. This lesion-mimic phenotype suggests that constitutive degradation of MLG could somehow trigger a signalling pathway leading to constitutive activation of plant defences. Recent reports demonstrate that β-1,3-1,4-glucanase-derived oligosaccharides are recognized by plants and act as defence signalling molecules. Even exogenous applications of short MLG fragments have a protective effect against pathogen attack not only in monocot, but also dicot plant species [189,190]. Dicot plant walls do not contain MLG, raising the question of why these plants can recognize products of MLG degradation. Although the mechanism has not been elucidated, it has been suggested that dicot plants could use β-glucanases to partially degrade MLG present in the walls of pathogens. MLG hydrolysis might directly hinder the pathogen growth, or alternatively, the resulting (1,3;1,4)-β-D-glucooligosaccharides could act as pathogen-associated molecular patterns (PAMPs), whose detection would trigger the activation of downstream plant defences. The presence of MLG in plant pathogens is poorly characterized and has only been demonstrated in a few examples, such as in the fungi *Rhynchosporium secalis*, *Aspergillus fumigatus* and *Neurospora crassa*, the oomycete *Hyaloperonospera arabidopsidis* or the endosymbiotic bacterium *Ensifer meliloti* [190–194]. MLG oligosaccharide-triggered protection has been reported for pathogens with completely different lifestyles and infection strategies such as the necrotrophic pathogens *Sclerotinia sclerotiorum* and *Botrytis cinerea* or biotrophs such as the oomycete *Hyaloperonospora arabidopsidis* or the hemibiotrophic bacterium *Pseudomonas syringae*. Considering these results, it has been speculated that MLG might be a more widespread wall component in plant pathogens than previously thought [189,190]. The existence of such a mechanism could imply the presence of hitherto uncharacterized plant β-glucanases encoded in the genome of dicot species that generate MLG-derived signalling molecules [26,189]. Thus far, no β-1,3-1,4-glucanase has been identified in *Arabidopsis*. Alternatively, β-1,4-glucanases might be responsible for the production of immunogenic MLG oligos, as these enzymes hydrolyse internal β-(1,4)-glucosidic linkages in MLG, although the products of hydrolysis are different from those derived from β-1,3-1,4-glucanase action (Figure 1) [8,195]. Supporting that hypothesis, MLG oligosaccharide structures compatible with cellulase origin display similar defence-related responses than those derived from β-1,3-1,4-glucanase hydrolysis [190]. Unfortunately, only a small number of the β-glucanases from *Arabidopsis* or other dicot species have been biochemically characterized. Hence, the study of such enzymes represents an exciting task for the future.

## 5. Conclusions

Plant β-glucanases catalyse the hydrolysis of β-glucosidic linkages found in the structure of polysaccharides present in the cell wall of plants and microbes (Figure 1). This catalytic ability confers various physiological roles to plant growth, development and interaction with the environment (Table 1). Despite recent advances in identifying new physiological functions, characterization of the precise biochemical activities is needed to decipher the importance of β-glucanases in plant adaptation towards environmental cues.

**Table 1.** Proposed physiological roles of plant β-glucanases.

| Type | Substrate | Physiological Roles |
|---|---|---|
| 1,4-β-glucanases | Crystalline cellulose | Irreversible wall disassembly: root hair emergence, endosperm breakdown, fruit ripening |
| | Amorphous cellulose | Secondary wall mechanical strength<br>Wall remodelling: cell expansion, fruit ripening, nematode attack<br>Cell-cell adhesion: grafting, plant parasitism<br>Cellulose biosynthesis (plasma membrane-associated) |
| | MLG [1]<br>(Fungal wall) | Antifungal activity<br>Elicitor release: MAMP [2] |
| 1,3-β-glucanases | Callose | Plasmodesmata and symplastic transport: dormancy release, fruit development, cell-to-cell communication<br>Reproductive organs: pollen, style and stigma development |
| | 1,3-β-glucan<br>(Fungal wall) | Antifungal activity<br>Elicitor release<br>Fungal effector release |
| 1,3-1,4-β-glucanases | MLG [1] | Cell wall loosening during germination<br>Energy source in the dark |
| | MLG [1]<br>(Fungal wall) | Antifungal activity<br>Elicitor release |

[1] MLG—mixed-linkage glucan; [2] MAMP—microbe-associated molecular patterns.

**Author Contributions:** Conceptualization, V.R. and M.P.; Visualization T.P., M.P. and V.R., Writing—Original Draft Preparation, T.P. and V.R.; Writing—Review and Editing, M.P. All authors have read and agreed to the published version of the manuscript.

**Funding:** This work has been funded by the Germany's Federal Ministry of Education and Research (BMBF) grant "Cornwall", 031B0193A and by the Deutsche Forschungsgemeinschaft (DFG, German Research Foundation) under Germany's Excellence Strategy—EXC 2048/1—Project ID: 390686111 to M.P. Additional funding was provided by Marie Curie PIOF-GA-2013-623553 to V.R., and DFG grant "MOPOST" Project ID: 410274474 funding T.P.

**Institutional Review Board Statement:** Not applicable.

**Informed Consent Statement:** Not applicable.

**Data Availability Statement:** Not applicable.

**Conflicts of Interest:** The authors declare no conflict of interest.

## References

1. Houston, K.; Tucker, M.R.; Chowdhury, J.; Shirley, N.; Little, A. The Plant Cell Wall: A Complex and Dynamic Structure As Revealed by the Responses of Genes under Stress Conditions. *Front. Plant Sci.* **2016**, *7*, 984. [CrossRef] [PubMed]
2. Zhang, B.; Gao, Y.; Zhang, L.; Zhou, Y. The Plant Cell Wall: Biosynthesis, Construction, and Functions. *J. Integr. Plant Biol.* **2021**, *63*, 251–272. [CrossRef] [PubMed]
3. Labavitch, J.M. Cell Wall Turnover in Plant Development. *Annu. Rev. Plant Physiol.* **1981**, *32*, 385–406. [CrossRef]
4. Tucker, M.R.; Lou, H.; Aubert, M.K.; Wilkinson, L.G.; Little, A.; Houston, K.; Pinto, S.C.; Shirley, N.J. Exploring the Role of Cell Wall-Related Genes and Polysaccharides during Plant Development. *Plants* **2018**, *7*, 42. [CrossRef] [PubMed]
5. Cantarel, B.L.; Coutinho, P.M.; Rancurel, C.; Bernard, T.; Lombard, V.; Henrissat, B. The Carbohydrate-Active EnZymes Database (CAZy): An Expert Resource for Glycogenomics. *Nucleic Acids Res.* **2009**, *37*, D233–D238. [CrossRef]
6. Schomburg, I.; Chang, A.; Schomburg, D. BRENDA, Enzyme Data and Metabolic Information. *Nucleic Acids Res.* **2002**, *30*, 47–49. [CrossRef]
7. Simmons, C.R. The Physiology and Molecular Biology of Plant 1,3-β-D-Glucanases and 1,3;1,4-β-D-Glucanases. *Crit. Rev. Plant Sci.* **1994**, *13*, 325–387. [CrossRef]
8. Høj, P.B.; Fincher, G.B. Molecular Evolution of Plant Beta-Glucan Endohydrolases. *Plant J.* **1995**, *7*, 367–379. [CrossRef]
9. Del Campillo, E. Multiple Endo-1,4-Beta-D-Glucanase (Cellulase) Genes in *Arabidopsis*. *Curr. Top. Dev. Biol.* **1999**, *46*, 39–61. [CrossRef]
10. Henrissat, B. A Classification of Glycosyl Hydrolases Based on Amino Acid Sequence Similarities. *Biochem. J.* **1991**, *280 Pt 2*, 309–316. [CrossRef]
11. Wilson, D.; Urbanowicz, B. Glycoside Hydrolase Family 9/Plant Endoglucanases. Available online: https://www.cazypedia.org/index.php/Glycoside_Hydrolase_Family_9/Plant_endoglucanases (accessed on 22 March 2022).
12. Nicol, F.; His, I.; Jauneau, A.; Vernhettes, S.; Canut, H.; Höfte, H. A Plasma Membrane-Bound Putative Endo-1,4-Beta-D-Glucanase Is Required for Normal Wall Assembly and Cell Elongation in Arabidopsis. *EMBO J.* **1998**, *17*, 5563–5576. [CrossRef] [PubMed]
13. Szyjanowicz, P.M.J.; McKinnon, I.; Taylor, N.G.; Gardiner, J.; Jarvis, M.C.; Turner, S.R. The Irregular Xylem 2 Mutant Is an Allele of Korrigan That Affects the Secondary Cell Wall of *Arabidopsis thaliana*. *Plant J.* **2004**, *37*, 730–740. [CrossRef] [PubMed]
14. Vain, T.; Crowell, E.F.; Timpano, H.; Biot, E.; Desprez, T.; Mansoori, N.; Trindade, L.M.; Pagant, S.; Robert, S.; Höfte, H.; et al. The Cellulase KORRIGAN Is Part of the Cellulose Synthase Complex. *Plant Physiol.* **2014**, *165*, 1521–1532. [CrossRef] [PubMed]
15. Eklöf, J.M.; Brumer, H. The XTH Gene Family: An Update on Enzyme Structure, Function, and Phylogeny in Xyloglucan Remodeling. *Plant Physiol.* **2010**, *153*, 456–466. [CrossRef]
16. Smith, R.C.; Fry, S.C. Endotransglycosylation of Xyloglucans in Plant Cell Suspension Cultures. *Biochem. J.* **1991**, *279*, 529–535. [CrossRef]
17. Nishitani, K.; Tominaga, R. Endo-Xyloglucan Transferase, a Novel Class of Glycosyltransferase That Catalyzes Transfer of a Segment of Xyloglucan Molecule to Another Xyloglucan Molecule. *J. Biol. Chem.* **1992**, *267*, 21058–21064. [CrossRef]
18. Rose, J.K.C.; Braam, J.; Fry, S.C.; Nishitani, K. The XTH Family of Enzymes Involved in Xyloglucan Endotransglucosylation and Endohydrolysis: Current Perspectives and a New Unifying Nomenclature. *Plant Cell Physiol.* **2002**, *43*, 1421–1435. [CrossRef]
19. Herburger, K.; Franková, L.; Pičmanová, M.; Loh, J.W.; Valenzuela-Ortega, M.; Meulewaeter, F.; Hudson, A.D.; French, C.E.; Fry, S.C. Hetero-Trans-β-Glucanase Produces Cellulose-Xyloglucan Covalent Bonds in the Cell Walls of Structural Plant Tissues and Is Stimulated by Expansin. *Mol. Plant* **2020**, *13*, 1047–1062. [CrossRef]
20. Ishida, K.; Yokoyama, R. Reconsidering the Function of the Xyloglucan Endotransglucosylase/Hydrolase Family. *J. Plant Res.* **2022**, *135*, 145–156. [CrossRef]
21. Hrmova, M.; Stratilová, B.; Stratilová, E. Broad Specific Xyloglucan:Xyloglucosyl Transferases Are Formidable Players in the Re-Modelling of Plant Cell Wall Structures. *Int. J. Mol. Sci.* **2022**, *23*, 1656. [CrossRef]
22. Moore, A.E.; Stone, B.A. A β-1,3-Glucan Hydrolase from Nicotiana Glutinosa II. Specificity, Action Pattern and Inhibitor Studies. *Biochim. Biophys. Acta BBA-Enzymol.* **1972**, *258*, 248–264. [CrossRef]
23. Bowman, S.M.; Free, S.J. The Structure and Synthesis of the Fungal Cell Wall. *BioEssays* **2006**, *28*, 799–808. [CrossRef]
24. Balasubramanian, V.; Vashisht, D.; Cletus, J.; Sakthivel, N. Plant β-1,3-Glucanases: Their Biological Functions and Transgenic Expression against Phytopathogenic Fungi. *Biotechnol. Lett.* **2012**, *34*, 1983–1990. [CrossRef]
25. Abeles, F.B.; Bosshart, R.P.; Forrence, L.E.; Habig, W.H. Preparation and Purification of Glucanase and Chitinase from Bean Leaves 1. *Plant Physiol.* **1971**, *47*, 129–134. [CrossRef]
26. Lombard, V.; Ramulu, H.G.; Drula, E.; Coutinho, P.M.; Henrissat, B. The Carbohydrate-Active Enzymes Database (CAZy) in 2013. *Nucleic Acids Res.* **2014**, *42*, D490–D495. [CrossRef]

27. Parrish, F.W.; Perlin, A.S.; Reese, E.T. Selective Enzymolysis of Poly-β-D-glucans, and The Structure of The Polymers. *Can. J. Chem.* **1960**, *38*, 2094–2104. [CrossRef]
28. McCleary, B.V. Purification of (1 → 3),(1 → 4)-β-d-Glucan from Barley Flour. In *Methods in Enzymology*; Biomass Part A: Cellulose and Hemicellulose; Academic Press: Cambridge, MA, USA, 1988; Volume 160, pp. 511–514.
29. Perlin, A.S.; Suzuki, S. The Structure of Lichenin: Selective Enzymolysis Studies. *Can. J. Chem.* **1962**, *40*, 50–56. [CrossRef]
30. Olafsdottir, E.S.; Ingólfsdottir, K. Polysaccharides from Lichens: Structural Characteristics and Biological Activity. *Planta Med.* **2001**, *67*, 199–208. [CrossRef]
31. Stone, B.A. Chemistry of β-Glucans. In *Chemistry, Biochemistry, and Biology of 1–3 Beta Glucans and Related Polysaccharides*; Bacic, A., Fincher, G.B., Stone, B.A., Eds.; Academic Press: San Diego, CA, USA, 2009; Chapter 2.1; pp. 5–46, ISBN 978-0-12-373971-1.
32. Varghese, J.N.; Garrett, T.P.; Colman, P.M.; Chen, L.; Høj, P.B.; Fincher, G.B. Three-Dimensional Structures of Two Plant Beta-Glucan Endohydrolases with Distinct Substrate Specificities. *Proc. Natl. Acad. Sci. USA* **1994**, *91*, 2785–2789. [CrossRef]
33. Bulone, V.; Schwerdt, J.G.; Fincher, G.B. Co-Evolution of Enzymes Involved in Plant Cell Wall Metabolism in the Grasses. *Front. Plant Sci.* **2019**, *10*, 1009. [CrossRef]
34. Libertini, E.; Li, Y.; McQueen-Mason, S.J. Phylogenetic Analysis of the Plant Endo-Beta-1,4-Glucanase Gene Family. *J. Mol. Evol.* **2004**, *58*, 506–515. [CrossRef]
35. Urbanowicz, B.R.; Bennett, A.B.; del Campillo, E.; Catalá, C.; Hayashi, T.; Henrissat, B.; Höfte, H.; McQueen-Mason, S.J.; Patterson, S.E.; Shoseyov, O.; et al. Structural Organization and a Standardized Nomenclature for Plant Endo-1,4-β-Glucanases (Cellulases) of Glycosyl Hydrolase Family 9. *Plant Physiol.* **2007**, *144*, 1693–1696. [CrossRef]
36. Buchanan, M.; Burton, R.A.; Dhugga, K.S.; Rafalski, A.J.; Tingey, S.V.; Shirley, N.J.; Fincher, G.B. Endo-(1,4)-β-Glucanase Gene Families in the Grasses: Temporal and Spatial Co-Transcription of Orthologous Genes1. *BMC Plant Biol.* **2012**, *12*, 235. [CrossRef]
37. Du, Q.; Wang, L.; Yang, X.; Gong, C.; Zhang, D. Populus Endo-β-1,4-Glucanases Gene Family: Genomic Organization, Phylogenetic Analysis, Expression Profiles and Association Mapping. *Planta* **2015**, *241*, 1417–1434. [CrossRef]
38. Henrissat, B.; Claeyssens, M.; Tomme, P.; Lemesle, L.; Mornon, J.-P. Cellulase Families Revealed by Hydrophobic Cluster Analysi. *Gene* **1989**, *81*, 83–95. [CrossRef]
39. Davies, G.; Henrissat, B. Structures and Mechanisms of Glycosyl Hydrolases. *Structure* **1995**, *3*, 853–859. [CrossRef]
40. Jumper, J.; Evans, R.; Pritzel, A.; Green, T.; Figurnov, M.; Ronneberger, O.; Tunyasuvunakool, K.; Bates, R.; Žídek, A.; Potapenko, A.; et al. Highly Accurate Protein Structure Prediction with AlphaFold. *Nature* **2021**, *596*, 583–589. [CrossRef]
41. Varadi, M.; Anyango, S.; Deshpande, M.; Nair, S.; Natassia, C.; Yordanova, G.; Yuan, D.; Stroe, O.; Wood, G.; Laydon, A.; et al. AlphaFold Protein Structure Database: Massively Expanding the Structural Coverage of Protein-Sequence Space with High-Accuracy Models. *Nucleic Acids Res.* **2022**, *50*, D439–D444. [CrossRef]
42. Kundu, S.; Sharma, R. Origin, Evolution, and Divergence of Plant Class C GH9 Endoglucanases. *BMC Evol. Biol.* **2018**, *18*, 79. [CrossRef]
43. Kundu, S. Insights into the Mechanism(s) of Digestion of Crystalline Cellulose by Plant Class C GH9 Endoglucanases. *J. Mol. Model.* **2019**, *25*, 240. [CrossRef]
44. Mølhøj, M.; Johansen, B.; Ulvskov, P.; Borkhardt, B. Expression of a Membrane-Anchored Endo-1,4-β-Glucanase from *Brassica napus*, Orthologous to KOR from *Arabidopsis thaliana*, Is Inversely Correlated to Elongation in Light-Grown Plants. *Plant Mol. Biol.* **2001**, *45*, 93–105. [CrossRef] [PubMed]
45. Mølhøj, M.; Pagant, S.; Höfte, H. Towards Understanding the Role of Membrane-Bound Endo-β-1,4-Glucanases in Cellulose Biosynthesis. *Plant Cell Physiol.* **2002**, *43*, 1399–1406. [CrossRef] [PubMed]
46. Gu, Y.; Rasmussen, C.G. Cell Biology of Primary Cell Wall Synthesis in Plants. *Plant Cell* **2022**, *34*, 103–128. [CrossRef] [PubMed]
47. Mansoori, N.; Timmers, J.; Desprez, T.; Kamei, C.L.A.; Dees, D.C.T.; Vincken, J.-P.; Visser, R.G.F.; Höfte, H.; Vernhettes, S.; Trindade, L.M. KORRIGAN1 Interacts Specifically with Integral Components of the Cellulose Synthase Machinery. *PLoS ONE* **2014**, *9*, e112387. [CrossRef]
48. Worden, N.; Wilkop, T.E.; Esteve, V.E.; Jeannotte, R.; Lathe, R.; Vernhettes, S.; Weimer, B.; Hicks, G.; Alonso, J.; Labavitch, J.; et al. CESA TRAFFICKING INHIBITOR Inhibits Cellulose Deposition and Interferes with the Trafficking of Cellulose Synthase Complexes and Their Associated Proteins KORRIGAN1 and POM2/CELLULOSE SYNTHASE INTERACTIVE PROTEIN1. *Plant Physiol.* **2015**, *167*, 381–393. [CrossRef]
49. Zhang, Y.; Nikolovski, N.; Sorieul, M.; Vellosillo, T.; McFarlane, H.E.; Dupree, R.; Kesten, C.; Schneider, R.; Driemeier, C.; Lathe, R.; et al. Golgi-Localized STELLO Proteins Regulate the Assembly and Trafficking of Cellulose Synthase Complexes in *Arabidopsis*. *Nat. Commun.* **2016**, *7*, 11656. [CrossRef]
50. Takahashi, J.; Rudsander, U.J.; Hedenström, M.; Banasiak, A.; Harholt, J.; Amelot, N.; Immerzeel, P.; Ryden, P.; Endo, S.; Ibatullin, F.M.; et al. *KORRIGAN1* and Its Aspen Homolog *PttCel9A1* Decrease Cellulose Crystallinity in Arabidopsis Stems. *Plant Cell Physiol.* **2009**, *50*, 1099–1115. [CrossRef]
51. Peng, L.; Kawagoe, Y.; Hogan, P.; Delmer, D. Sitosterol-β-Glucoside as Primer for Cellulose Synthesis in Plants. *Science* **2002**, *295*, 147–150. [CrossRef]
52. Somerville, C. Cellulose Synthesis in Higher Plants. *Annu. Rev. Cell Dev. Biol.* **2006**, *22*, 53–78. [CrossRef]
53. Bhandari, S.; Fujino, T.; Thammanagowda, S.; Zhang, D.; Xu, F.; Joshi, C.P. Xylem-Specific and Tension Stress-Responsive Coexpression of KORRIGAN Endoglucanase and Three Secondary Wall-Associated Cellulose Synthase Genes in Aspen Trees. *Planta* **2006**, *224*, 828–837. [CrossRef]

54. Zhou, H.-L.; He, S.-J.; Cao, Y.-R.; Chen, T.; Du, B.-X.; Chu, C.-C.; Zhang, J.-S.; Chen, S.-Y. OsGLU1, a Putative Membrane-Bound Endo-1,4-Beta-D-Glucanase from Rice, Affects Plant Internode Elongation. *Plant Mol. Biol.* **2006**, *60*, 137–151. [CrossRef] [PubMed]
55. Kalluri, U.C.; Payyavula, R.S.; Labbé, J.L.; Engle, N.; Bali, G.; Jawdy, S.S.; Sykes, R.W.; Davis, M.; Ragauskas, A.; Tuskan, G.A.; et al. Down-Regulation of KORRIGAN-Like Endo-β-1,4-Glucanase Genes Impacts Carbon Partitioning, Mycorrhizal Colonization and Biomass Production in Populus. *Front. Plant Sci.* **2016**, *7*, 1455. [CrossRef] [PubMed]
56. Brummell, D.A.; Catala, C.; Lashbrook, C.C.; Bennett, A.B. A Membrane-Anchored E-Type Endo-1,4-β-Glucanase Is Localized on Golgi and Plasma Membranes of Higher Plants. *Proc. Natl. Acad. Sci. USA* **1997**, *94*, 4794–4799. [CrossRef] [PubMed]
57. Yu, L.; Sun, J.; Li, L. PtrCel9A6, an Endo-1,4-β-Glucanase, Is Required for Cell Wall Formation during Xylem Differentiation in Populus. *Mol. Plant* **2013**, *6*, 1904–1917. [CrossRef] [PubMed]
58. Yu, L.; Chen, H.; Sun, J.; Li, L. PtrKOR1 Is Required for Secondary Cell Wall Cellulose Biosynthesis in Populus. *Tree Physiol.* **2014**, *34*, 1289–1300. [CrossRef]
59. López-Cruz, J.; Finiti, I.; Fernández-Crespo, E.; Crespo-Salvador, O.; García-Agustín, P.; González-Bosch, C. Absence of Endo-1,4-β-Glucanase KOR1 Alters the Jasmonate-Dependent Defence Response to *Pseudomonas Syringae* in *Arabidopsis*. *J. Plant Physiol.* **2014**, *171*, 1524–1532. [CrossRef]
60. Rose, J.K.C.; Bennett, A.B. Cooperative Disassembly of the Cellulose-Xyloglucan Network of Plant Cell Walls: Parallels between Cell Expansion and Fruit Ripening. *Trends Plant Sci.* **1999**, *4*, 176–183. [CrossRef]
61. Brummell, D.A.; Hall, B.D.; Bennett, A.B. Antisense Suppression of Tomato Endo-1,4-Beta-Glucanase Cel2 MRNA Accumulation Increases the Force Required to Break Fruit Abscission Zones but Does Not Affect Fruit Softening. *Plant Mol. Biol.* **1999**, *40*, 615–622. [CrossRef]
62. Cosgrove, D.J. Growth of the Plant Cell Wall. *Nat. Rev. Mol. Cell Biol.* **2005**, *6*, 850–861. [CrossRef]
63. Del Campillo, E.; Abdel-Aziz, A.; Crawford, D.; Patterson, S.E. Root Cap Specific Expression of an Endo-β-1,4-Glucanase (Cellulase): A New Marker to Study Root Development in *Arabidopsis*. *Plant Mol. Biol.* **2004**, *56*, 309–323. [CrossRef]
64. Lewis, D.R.; Olex, A.L.; Lundy, S.R.; Turkett, W.H.; Fetrow, J.S.; Muday, G.K. A Kinetic Analysis of the Auxin Transcriptome Reveals Cell Wall Remodeling Proteins That Modulate Lateral Root Development in *Arabidopsis*. *Plant Cell* **2013**, *25*, 3329–3346. [CrossRef] [PubMed]
65. Karve, R.; Suárez-Román, F.; Iyer-Pascuzzi, A.S. The Transcription Factor NIN-LIKE PROTEIN7 Controls Border-Like Cell Release. *Plant Physiol.* **2016**, *171*, 2101–2111. [CrossRef] [PubMed]
66. Fischer, R.L.; Bennett, A.B. Role of Cell Wall Hydrolases in Fruit Ripening. *Annu. Rev. Plant Physiol. Plant Mol. Biol.* **1991**, *42*, 675–703. [CrossRef]
67. Harpster, M.H.; Lee, K.Y.; Dunsmuir, P. Isolation and Characterization of a Gene Encoding Endo-β-1,4-Glucanase from Pepper (*Capsicum annuum* L.). *Plant Mol. Biol.* **1997**, *33*, 47–59. [CrossRef]
68. Manning, K. Isolation of a Set of Ripening-Related Genes from Strawberry: Their Identification and Possible Relationship to Fruit Quality Traits. *Planta* **1998**, *205*, 622–631. [CrossRef]
69. Jara, K.; Castro, R.I.; Ramos, P.; Parra-Palma, C.; Valenzuela-Riffo, F.; Morales-Quintana, L. Molecular Insights into FaEG1, a Strawberry Endoglucanase Enzyme Expressed during Strawberry Fruit Ripening. *Plants* **2019**, *8*, 140. [CrossRef]
70. Ohmiya, Y.; Takeda, T.; Nakamura, S.; Sakai, F.; Hayashi, T. Purification and Properties of a Wall-Bound Endo-1,4-β-Glucanase from Suspension-Cultured Poplar Cells1. *Plant Cell Physiol.* **1995**, *36*, 607–614. [CrossRef]
71. Ohmiya, Y.; Nakai, T.; Park, Y.W.; Aoyama, T.; Oka, A.; Sakai, F.; Hayashi, T. The Role of PopCel1 and PopCel2 in Poplar Leaf Growth and Cellulose Biosynthesis. *Plant J.* **2003**, *33*, 1087–1097. [CrossRef]
72. Park, Y.W.; Tominaga, R.; Sugiyama, J.; Furuta, Y.; Tanimoto, E.; Samejima, M.; Sakai, F.; Hayashi, T. Enhancement of Growth by Expression of Poplar Cellulase in *Arabidopsis thaliana*. *Plant J.* **2003**, *33*, 1099–1106. [CrossRef]
73. Huang, J.; Xia, T.; Li, G.; Li, X.; Li, Y.; Wang, Y.; Wang, Y.; Chen, Y.; Xie, G.; Bai, F.-W.; et al. Overproduction of Native Endo-β-1,4-Glucanases Leads to Largely Enhanced Biomass Saccharification and Bioethanol Production by Specific Modification of Cellulose Features in Transgenic Rice. *Biotechnol. Biofuels* **2019**, *12*, 11. [CrossRef]
74. Glass, M.; Barkwill, S.; Unda, F.; Mansfield, S.D. Endo-β-1,4-Glucanases Impact Plant Cell Wall Development by Influencing Cellulose Crystallization. *J. Integr. Plant Biol.* **2015**, *57*, 396–410. [CrossRef] [PubMed]
75. Schwacke, R.; Schneider, A.; van der Graaff, E.; Fischer, K.; Catoni, E.; Desimone, M.; Frommer, W.B.; Flügge, U.-I.; Kunze, R. ARAMEMNON, a Novel Database for Arabidopsis Integral Membrane Proteins. *Plant Physiol.* **2003**, *131*, 16–26. [CrossRef] [PubMed]
76. Song, D.; Xi, W.; Shen, J.; Bi, T.; Li, L. Characterization of the Plasma Membrane Proteins and Receptor-like Kinases Associated with Secondary Vascular Differentiation in Poplar. *Plant Mol. Biol.* **2011**, *76*, 97–115. [CrossRef] [PubMed]
77. Nilsson, R.; Bernfur, K.; Gustavsson, N.; Bygdell, J.; Wingsle, G.; Larsson, C. Proteomics of Plasma Membranes from Poplar Trees Reveals Tissue Distribution of Transporters, Receptors, and Proteins in Cell Wall Formation *. *Mol. Cell. Proteom.* **2010**, *9*, 368–387. [CrossRef]
78. Wang, J.; Jiang, L.; Wu, R. Plant Grafting: How Genetic Exchange Promotes Vascular Reconnection. *New Phytol.* **2017**, *214*, 56–65. [CrossRef]
79. Notaguchi, M.; Kurotani, K.; Sato, Y.; Tabata, R.; Kawakatsu, Y.; Okayasu, K.; Sawai, Y.; Okada, R.; Asahina, M.; Ichihashi, Y.; et al. Cell-Cell Adhesion in Plant Grafting Is Facilitated by β-1,4-Glucanases. *Science* **2020**, *369*, 698–702. [CrossRef]

80. Kurotani, K.; Wakatake, T.; Ichihashi, Y.; Okayasu, K.; Sawai, Y.; Ogawa, S.; Cui, S.; Suzuki, T.; Shirasu, K.; Notaguchi, M. Host-Parasite Tissue Adhesion by a Secreted Type of β-1,4-Glucanase in the Parasitic Plant *Phtheirospermum Japonicum*. *Commun. Biol.* **2020**, *3*, 407. [CrossRef]
81. Finiti, I.; Leyva, M.O.; López-Cruz, J.; Rodrigues, B.C.; Vicedo, B.; Angulo, C.; Bennett, A.B.; Grant, M.; García-Agustín, P.; González-Bosch, C. Functional Analysis of Endo-1,4-β-Glucanases in Response to *Botrytis Cinerea* and *Pseudomonas Syringae* Reveals Their Involvement in Plant–Pathogen Interactions. *Plant Biol.* **2013**, *15*, 819–831. [CrossRef]
82. Flors, V.; Leyva, M.D.L.O.; Vicedo, B.; Finiti, I.; Real, M.D.; García-Agustín, P.; Bennett, A.B.; González-Bosch, C. Absence of the Endo-β-1,4-Glucanases Cel1 and Cel2 Reduces Susceptibility to *Botrytis Cinerea* in Tomato. *Plant J.* **2007**, *52*, 1027–1040. [CrossRef]
83. Woo, M.-O.; Beard, H.; MacDonald, M.H.; Brewer, E.P.; Youssef, R.M.; Kim, H.; Matthews, B.F. Manipulation of Two α-Endo-β-1,4-Glucanase Genes, *AtCel6* and *GmCel7*, Reduces Susceptibility to *Heterodera Glycines* in Soybean Roots. *Mol. Plant Pathol.* **2014**, *15*, 927–939. [CrossRef]
84. Goellner, M.; Wang, X.; Davis, E.L. Endo-β-1,4-Glucanase Expression in Compatible Plant-Nematode Interactions. *Plant Cell* **2001**, *13*, 2241–2256. [CrossRef] [PubMed]
85. Endo, B.Y. Histology and Ultrastructural Modification Induced By Cyst Nematodes. In *Cyst Nematodes*; Lamberti, F., Taylor, C.E., Eds.; Nato ASI Series; Springer: Boston, MA, USA, 1986; pp. 133–146, ISBN 978-1-4613-2251-1.
86. Hussey, R.; Grundler, F. Nematode Parasitism of Plants. In *Physiology and Biochemistry of Free-Living and Plant Parasitic Nematodes*; CAB International Press: Oxfordshire, UK, 1998; pp. 213–243.
87. Tucker, M.L.; Burke, A.; Murphy, C.A.; Thai, V.K.; Ehrenfried, M.L. Gene Expression Profiles for Cell Wall-Modifying Proteins Associated with Soybean Cyst Nematode Infection, Petiole Abscission, Root Tips, Flowers, Apical Buds, and Leaves. *J. Exp. Bot.* **2007**, *58*, 3395–3406. [CrossRef] [PubMed]
88. Wieczorek, K.; Hofmann, J.; Blöchl, A.; Szakasits, D.; Bohlmann, H.; Grundler, F.M.W. Arabidopsis Endo-1,4-β-Glucanases Are Involved in the Formation of Root Syncytia Induced by *Heterodera Schachtii*. *Plant J.* **2008**, *53*, 336–351. [CrossRef] [PubMed]
89. Matthews, B.; Ibrahim, H.; Klink, V. Changes in the Expression of Genes in Soybean Roots Infected by Nematodes. In *Soybean-Genetics and Novel Techniques for Yield Enhancement*; InTech: London, UK, 2011; ISBN 978-953-307-721-5.
90. Trainotti, L.; Spolaore, S.; Pavanello, A.; Baldan, B.; Casadoro, G. A Novel E-Type Endo-β-1,4-Glucanase with a Putative Cellulose-Binding Domain Is Highly Expressed in Ripening Strawberry Fruits. *Plant Mol. Biol.* **1999**, *40*, 323–332. [CrossRef]
91. Del Campillo, E.; Gaddam, S.; Mettle-Amuah, D.; Heneks, J. A Tale of Two Tissues: *AtGH9C1* Is an Endo-β-1,4-Glucanase Involved in Root Hair and Endosperm Development in *Arabidopsis*. *PLoS ONE* **2012**, *7*, e49363. [CrossRef]
92. Urbanowicz, B.R.; Catalá, C.; Irwin, D.; Wilson, D.B.; Ripoll, D.R.; Rose, J.K.C. A Tomato Endo-β-1,4-Glucanase, SlCel9C1, Represents a Distinct Subclass with a New Family of Carbohydrate Binding Modules (CBM49) *. *J. Biol. Chem.* **2007**, *282*, 12066–12074. [CrossRef]
93. McLean, B.W.; Bray, M.R.; Boraston, A.B.; Gilkes, N.R.; Haynes, C.A.; Kilburn, D.G. Analysis of Binding of the Family 2a Carbohydrate-Binding Module from Cellulomonas Fimi Xylanase 10A to Cellulose: Specificity and Identification of Functionally Important Amino Acid Residues. *Protein Eng.* **2000**, *13*, 801–809. [CrossRef]
94. Duan, C.-J.; Feng, Y.-L.; Cao, Q.-L.; Huang, M.-Y.; Feng, J.-X. Identification of a Novel Family of Carbohydrate-Binding Modules with Broad Ligand Specificity. *Sci. Rep.* **2016**, *6*, 19392. [CrossRef]
95. Boraston, A.B.; Nurizzo, D.; Notenboom, V.; Ducros, V.; Rose, D.R.; Kilburn, D.G.; Davies, G.J. Differential Oligosaccharide Recognition by Evolutionarily-Related β-1,4 and β-1,3 Glucan-Binding Modules. *J. Mol. Biol.* **2002**, *319*, 1143–1156. [CrossRef]
96. Charnock, S.J.; Bolam, D.N.; Nurizzo, D.; Szabó, L.; McKie, V.A.; Gilbert, H.J.; Davies, G.J. Promiscuity in Ligand-Binding: The Three-Dimensional Structure of a *Piromyces* Carbohydrate-Binding Module, CBM29-2, in Complex with Cello- and Mannohexaose. *Proc. Natl. Acad. Sci. USA* **2002**, *99*, 14077–14082. [CrossRef]
97. Kim, D.Y.; Lee, M.J.; Cho, H.-Y.; Lee, J.S.; Lee, M.-H.; Chung, C.W.; Shin, D.-H.; Rhee, Y.H.; Son, K.-H.; Park, H.-Y. Genetic and Functional Characterization of an Extracellular Modular GH6 Endo-β-1,4-Glucanase from an Earthworm Symbiont, *Cellulosimicrobium Funkei* HY-13. *Antonie Van Leeuwenhoek* **2016**, *109*, 1–12. [CrossRef] [PubMed]
98. Yoshida, K.; Komae, K. A Rice Family 9 Glycoside Hydrolase Isozyme with Broad Substrate Specificity for Hemicelluloses in Type II Cell Walls. *Plant Cell Physiol.* **2006**, *47*, 1541–1554. [CrossRef] [PubMed]
99. Henrissat, B.; Bairoch, A. New Families in the Classification of Glycosyl Hydrolases Based on Amino Acid Sequence Similarities. *Biochem. J.* **1993**, *293*, 781–788. [CrossRef] [PubMed]
100. Linthorst, H.J.; Melchers, L.S.; Mayer, A.; van Roekel, J.S.; Cornelissen, B.J.; Bol, J.F. Analysis of Gene Families Encoding Acidic and Basic Beta-1,3-Glucanases of Tobacco. *Proc. Natl. Acad. Sci. USA* **1990**, *87*, 8756–8760. [CrossRef]
101. Hudspeth, R.L.; Hobbs, S.L.; Anderson, D.M.; Grula, J.W. Characterization and Expression of Chitinase and 1,3-β-Glucanase Genes in Cotton. *Plant Mol. Biol.* **1996**, *31*, 911–916. [CrossRef]
102. Jin, W.; Horner, H.T.; Palmer, R.G.; Shoemaker, R.C. Analysis and Mapping of Gene Families Encoding Beta-1,3-Glucanases of Soybean. *Genetics* **1999**, *153*, 445–452. [CrossRef]
103. Romero, G.O.; Simmons, C.; Yaneshita, M.; Doan, M.; Thomas, B.R.; Rodriguez, R.L. Characterization of Rice Endo-Beta-Glucanase Genes (*Gns2–Gns14*) Defines a New Subgroup within the Gene Family. *Gene* **1998**, *223*, 311–320. [CrossRef]
104. Hwang, D.H.; Kim, S.T.; Kim, S.G.; Kang, K.Y. Comprehensive Analysis of the Expression of Twenty-Seven Beta-1,3-Glucanase Genes in Rice (*Oryza sativa* L.). *Mol. Cells* **2007**, *23*, 207–214.

105. Doxey, A.C.; Yaish, M.W.F.; Moffatt, B.A.; Griffith, M.; McConkey, B.J. Functional Divergence in the *Arabidopsis* β-1,3-Glucanase Gene Family Inferred by Phylogenetic Reconstruction of Expression States. *Mol. Biol. Evol.* **2007**, *24*, 1045–1055. [CrossRef]
106. Xu, X.; Feng, Y.; Fang, S.; Xu, J.; Wang, X.; Guo, W. Genome-Wide Characterization of the β-1,3-Glucanase Gene Family in *Gossypium* by Comparative Analysis. *Sci. Rep.* **2016**, *6*, 29044. [CrossRef]
107. Barral, P.; Suárez, C.; Batanero, E.; Alfonso, C.; de Dios Alché, J.; Rodríguez-García, M.I.; Villalba, M.; Rivas, G.; Rodríguez, R. An Olive Pollen Protein with Allergenic Activity, Ole e 10, Defines a Novel Family of Carbohydrate-Binding Modules and Is Potentially Implicated in Pollen Germination. *Biochem. J.* **2005**, *390*, 77–84. [CrossRef] [PubMed]
108. Sticher, L.; Hinz, U.; Meyer, A.D.; Meins, F. Intracellular Transport and Processing of a Tobacco Vacuolar β-1,3-Glucanase. *Planta* **1992**, *188*, 559–565. [CrossRef] [PubMed]
109. Henrissat, B.; Davies, G.J. Glycoside Hydrolases and Glycosyltransferases. Families, Modules, and Implications for Genomics1. *Plant Physiol.* **2000**, *124*, 1515–1519. [CrossRef] [PubMed]
110. Leubner-Metzger, G.; Meins, F. Functions and Regulation of Plant SS-1,3-Glucanases (PR-2): Pathogenesis-Related Proteins in Plants. In *Pathogenesis-Related Proteins in Plants*; Datta, S.K., Muthukrishnan, S., Eds.; CRC Press, LLC: Boca Raton, FL, USA, 1999; pp. 49–76.
111. Bachman, E.S.; McClay, D.R. Molecular Cloning of the First Metazoan Beta-1,3 Glucanase from Eggs of the Sea Urchin *Strongylocentrotus Purpuratus*. *Proc. Natl. Acad. Sci. USA* **1996**, *93*, 6808–6813. [CrossRef]
112. Sun, L.; Gurnon, J.R.; Adams, B.J.; Graves, M.V.; Van Etten, J.L. Characterization of a β-1,3-Glucanase Encoded by Chlorella Virus PBCV-1. *Virology* **2000**, *276*, 27–36. [CrossRef]
113. Ruiz-Herrera, J.; Ortiz-Castellanos, L. Cell Wall Glucans of Fungi. A Review. *Cell Surf.* **2019**, *5*, 100022. [CrossRef]
114. Sela-Buurlage, M.B.; Ponstein, A.S.; Bres-Vloemans, S.A.; Melchers, L.S.; Van Den Elzen, P.J.M.; Cornelissen, B.J.C. Only Specific Tobacco (*Nicotiana tabacum*) Chitinases and [Beta]-1,3-Glucanases Exhibit Antifungal Activity. *Plant Physiol.* **1993**, *101*, 857–863. [CrossRef]
115. Amian, A.; Papenbrock, J.; Jacobsen, H.-J.; Hassan, F. Enhancing Transgenic Pea (*Pisum sativum* L.) Resistance against Fungal Diseases through Stacking of Two Antifungal Genes (Chitinase and Glucanase). *GM Crops* **2011**, *2*, 104–109. [CrossRef]
116. Zhang, S.-B.; Zhang, W.-J.; Zhai, H.-C.; Lv, Y.-Y.; Cai, J.-P.; Jia, F.; Wang, J.-S.; Hu, Y.-S. Expression of a Wheat β-1,3-Glucanase in *Pichia pastoris* and Its Inhibitory Effect on Fungi Commonly Associated with Wheat Kernel. *Protein Expr. Purif.* **2019**, *154*, 134–139. [CrossRef]
117. Liu, B.; Lu, Y.; Xin, Z.; Zhang, Z. Identification and Antifungal Assay of a Wheat β-1,3-Glucanase. *Biotechnol. Lett.* **2009**, *31*, 1005–1010. [CrossRef]
118. Wang, Y.; Liu, M.; Wang, X.; Zhong, L.; Shi, G.; Xu, Y.; Li, Y.; Li, R.; Huang, Y.; Ye, X.; et al. A Novel β-1,3-Glucanase Gns6 from Rice Possesses Antifungal Activity against *Magnaporthe oryzae*. *J. Plant Physiol.* **2021**, *265*, 153493. [CrossRef] [PubMed]
119. Gupta, P.; Ravi, I.; Sharma, V. Induction of β-1,3-Glucanase and Chitinase Activity in the Defense Response of *Eruca sativa* Plants against the Fungal Pathogen *Alternaria brassicicola*. *J. Plant Interact.* **2013**, *8*, 155–161. [CrossRef]
120. Salim, A.P.; Saminaidu, K.; Marimuthu, M.; Perumal, Y.; Rethinasamy, V.; Palanisami, J.R.; Vadivel, K. Defense Responses in Tomato Landrace and Wild Genotypes to Early Blight Pathogen *Alternaria solani* Infection and Accumulation of Pathogenesis-Related Proteins. *Arch. Phytopathol. Plant Prot.* **2011**, *44*, 1147–1164. [CrossRef]
121. Aggarwal, R.; Purwar, S.; Kharbikar, L.; Gupta, S. Induction of a Wheat β-1,3-Glucanase Gene during the Defense Response to *Bipolaris sorokiniana*. *Acta Phytopathol. Entomol. Hung.* **2011**, *46*, 39–47. [CrossRef]
122. Sundaresha, S.; Kumar, A.M.; Rohini, S.; Math, S.A.; Keshamma, E.; Chandrashekar, S.C.; Udayakumar, M. Enhanced Protection against Two Major Fungal Pathogens of Groundnut, *Cercospora arachidicola* and *Aspergillus flavus* in Transgenic Groundnut over-Expressing a Tobacco β 1–3 Glucanase. *Eur. J. Plant Pathol.* **2010**, *126*, 497–508. [CrossRef]
123. Wróbel-Kwiatkowska, M.; Lorenc-Kukula, K.; Starzycki, M.; Oszmiański, J.; Kepczyńska, E.; Szopa, J. Expression of β-1,3-Glucanase in Flax Causes Increased Resistance to Fungi. *Physiol. Mol. Plant Pathol.* **2004**, *65*, 245–256. [CrossRef]
124. Mackintosh, C.A.; Lewis, J.; Radmer, L.E.; Shin, S.; Heinen, S.J.; Smith, L.A.; Wyckoff, M.N.; Dill-Macky, R.; Evans, C.K.; Kravchenko, S.; et al. Overexpression of Defense Response Genes in Transgenic Wheat Enhances Resistance to Fusarium Head Blight. *Plant Cell Rep.* **2007**, *26*, 479–488. [CrossRef]
125. Maziah, M.; Sreeramanan, S.; Puad, A.; Meon, S. Production of Transgenic Banana Cultivar, Rastali (AAB) via *Agrobacterium*-Mediated Transformation with a Rice Chitinase Gene. *J. Plant Sci.* **2007**, *2*, 504–517. [CrossRef]
126. Fesel, P.H.; Zuccaro, A. β-Glucan: Crucial Component of the Fungal Cell Wall and Elusive MAMP in Plants. *Fungal Genet. Biol.* **2016**, *90*, 53–60. [CrossRef]
127. Yoshikawa, M.; Keen, N.T.; Wang, M.C. A Receptor on Soybean Membranes for a Fungal Elicitor of Phytoalexin Accumulation. *Plant Physiol.* **1983**, *73*, 497–506. [CrossRef]
128. Sharp, J.K.; Valent, B.; Albersheim, P. Purification and Partial Characterization of a Beta-Glucan Fragment That Elicits Phytoalexin Accumulation in Soybean. *J. Biol. Chem.* **1984**, *259*, 11312–11320. [CrossRef]
129. Yamaguchi, T.; Yamada, A.; Hong, N.; Ogawa, T.; Ishii, T.; Shibuya, N. Differences in the Recognition of Glucan Elicitor Signals between Rice and Soybean: β-Glucan Fragments from the Rice Blast Disease Fungus Pyricularia Oryzae That Elicit Phytoalexin Biosynthesis in Suspension-Cultured Rice Cells. *Plant Cell* **2000**, *12*, 817–826. [CrossRef] [PubMed]
130. Cosio, E.G.; Feger, M.; Miller, C.J.; Antelo, L.; Ebel, J. High-Affinity Binding of Fungal β-Glucan Elicitors to Cell Membranes of Species of the Plant Family Fabaceae. *Planta* **1996**, *200*, 92–99. [CrossRef]

131. Côté, F.; Roberts, K.A.; Hahn, M.G. Identification of High-Affinity Binding Sites for the Hepta-Beta-Glucoside Elicitor in Membranes of the Model Legumes *Medicago truncatula* and *Lotus japonicus*. *Planta* **2000**, *211*, 596–605. [CrossRef] [PubMed]
132. Wanke, A.; Malisic, M.; Wawra, S.; Zuccaro, A. Unraveling the Sugar Code: The Role of Microbial Extracellular Glycans in Plant-Microbe Interactions. *J. Exp. Bot.* **2021**, *72*, 15–35. [CrossRef] [PubMed]
133. Mélida, H.; Sopeña-Torres, S.; Bacete, L.; Garrido-Arandia, M.; Jordá, L.; López, G.; Muñoz-Barrios, A.; Pacios, L.F.; Molina, A. Non-Branched β-1,3-Glucan Oligosaccharides Trigger Immune Responses in Arabidopsis. *Plant J.* **2018**, *93*, 34–49. [CrossRef]
134. Wanke, A.; Rovenich, H.; Schwanke, F.; Velte, S.; Becker, S.; Hehemann, J.-H.; Wawra, S.; Zuccaro, A. Plant Species-Specific Recognition of Long and Short β-1,3-Linked Glucans Is Mediated by Different Receptor Systems. *Plant J.* **2020**, *102*, 1142–1156. [CrossRef]
135. Chandrasekar, B.; Wanke, A.; Wawra, S.; Saake, P.; Mahdi, L.; Charura, N.; Neidert, M.; Malisic, M.; Thiele, M.; Dama, M.; et al. Fungi Hijack a Ubiquitous Plant Apoplastic Endoglucanase to Release a ROS Scavenging β-Glucan Decasaccharide to Subvert Immune Responses. *Plant Cell*, 2022; *accepted*. [CrossRef]
136. Hird, D.L.; Worrall, D.; Hodge, R.; Smartt, S.; Paul, W.; Scott, R. The Anther-Specific Protein Encoded by the *Brassica Napus* and *Arabidopsis thaliana* A6 Gene Displays Similarity to β-1,3-Glucanases. *Plant J.* **1993**, *4*, 1023–1033. [CrossRef]
137. Parveen, S.; Mazumder, M.; Bhattacharya, A.; Mukhopadhyay, S.; Saha, U.; Mukherjee, A.; Mondal, B.; Debnath, A.J.; Das, S.; Sikdar, S.; et al. Identification of Anther-Specific Genes from Sesame and Functional Assessment of the Upstream Region of a Tapetum-Specific β-1,3-Glucanase Gene. *Plant Mol. Biol. Rep.* **2018**, *36*, 149–161. [CrossRef]
138. Zieliński, K.; Dubas, E.; Gerši, Z.; Krzewska, M.; Janas, A.; Nowicka, A.; Matušíková, I.; Żur, I.; Sakuda, S.; Moravčíková, J. β-1,3-Glucanases and Chitinases Participate in the Stress-Related Defence Mechanisms That Are Possibly Connected with Modulation of Arabinogalactan Proteins (AGP) Required for the Androgenesis Initiation in Rye (*Secale cereale* L.). *Plant Sci. Int. J. Exp. Plant Biol.* **2021**, *302*, 110700. [CrossRef] [PubMed]
139. Wan, L.; Zha, W.; Cheng, X.; Liu, C.; Lv, L.; Liu, C.; Wang, Z.; Du, B.; Chen, R.; Zhu, L.; et al. A Rice β-1,3-Glucanase Gene Osg1 Is Required for Callose Degradation in Pollen Development. *Planta* **2011**, *233*, 309–323. [CrossRef] [PubMed]
140. Delp, G.; Palva, E.T. A Novel Flower-Specific Arabidopsis Gene Related to Both Pathogen-Induced and Developmentally Regulated Plant Beta-1,3-Glucanase Genes. *Plant Mol. Biol.* **1999**, *39*, 565–575. [CrossRef] [PubMed]
141. De Storme, N.; Geelen, D. The Impact of Environmental Stress on Male Reproductive Development in Plants: Biological Processes and Molecular Mechanisms. *Plant Cell Environ.* **2014**, *37*, 1–18. [CrossRef] [PubMed]
142. Han, X.; Huang, L.-J.; Feng, D.; Jiang, W.; Miu, W.; Li, N. Plasmodesmata-Related Structural and Functional Proteins: The Long Sought-After Secrets of a Cytoplasmic Channel in Plant Cell Walls. *Int. J. Mol. Sci.* **2019**, *20*, 2946. [CrossRef] [PubMed]
143. Levy, A.; Erlanger, M.; Rosenthal, M.; Epel, B.L. A Plasmodesmata-Associated *β*-1,3-Glucanase in Arabidopsis. *Plant J.* **2007**, *49*, 669–682. [CrossRef]
144. Wu, S.-W.; Kumar, R.; Iswanto, A.B.B.; Kim, J.-Y. Callose Balancing at Plasmodesmata. *J. Exp. Bot.* **2018**, *69*, 5325–5339. [CrossRef]
145. Kim, I.; Zambryski, P.C. Cell-to-Cell Communication via Plasmodesmata during *Arabidopsis* Embryogenesis. *Curr. Opin. Plant Biol.* **2005**, *8*, 593–599. [CrossRef]
146. Benitez-Alfonso, Y.; Faulkner, C.; Pendle, A.; Miyashima, S.; Helariutta, Y.; Maule, A. Symplastic Intercellular Connectivity Regulates Lateral Root Patterning. *Dev. Cell* **2013**, *26*, 136–147. [CrossRef]
147. Guseman, J.M.; Lee, J.S.; Bogenschutz, N.L.; Peterson, K.M.; Virata, R.E.; Xie, B.; Kanaoka, M.M.; Hong, Z.; Torii, K.U. Dysregulation of Cell-to-Cell Connectivity and Stomatal Patterning by Loss-of-Function Mutation in *Arabidopsis Chorus* (*Glucan Synthase-like 8*). *Development* **2010**, *137*, 1731–1741. [CrossRef]
148. Gisel, A.; Hempel, F.D.; Barella, S.; Zambryski, P. Leaf-to-Shoot Apex Movement of Symplastic Tracer Is Restricted Coincident with Flowering in *Arabidopsis*. *Proc. Natl. Acad. Sci. USA* **2002**, *99*, 1713–1717. [CrossRef]
149. Paterlini, A.; Dorussen, D.; Fichtner, F.; van Rongen, M.; Delacruz, R.; Vojnović, A.; Helariutta, Y.; Leyser, O. Callose Accumulation in Specific Phloem Cell Types Reduces Axillary Bud Growth in *Arabidopsis thaliana*. *New Phytol.* **2021**, *231*, 516–523. [CrossRef]
150. Rinne, P.L.H.; Kaikuranta, P.M.; Van Der Schoot, C. The Shoot Apical Meristem Restores Its Symplasmic Organization during Chilling-Induced Release from Dormancy. *Plant J.* **2001**, *26*, 249–264. [CrossRef] [PubMed]
151. Gao, X.; Yuan, Y.; Liu, Z.; Liu, C.; Xin, H.; Zhang, Y.; Gai, S. Chilling and Gibberellin Acids Hyperinduce β-1,3-Glucanases to Reopen Transport Corridor and Break Endodormancy in Tree Peony (*Paeonia suffruticosa*). *Plant Physiol. Biochem.* **2021**, *167*, 771–784. [CrossRef] [PubMed]
152. Leubner-Metzger, G. Seed After-Ripening and over-Expression of Class I β-1,3-Glucanase Confer Maternal Effects on Tobacco Testa Rupture and Dormancy Release. *Planta* **2002**, *215*, 959–968. [CrossRef]
153. Choudhury, S.R.; Roy, S.; Sengupta, D.N. Characterization of Cultivar Differences in Beta-1,3 Glucanase Gene Expression, Glucanase Activity and Fruit Pulp Softening Rates during Fruit Ripening in Three Naturally Occurring Banana Cultivars. *Plant Cell Rep.* **2009**, *28*, 1641–1653. [CrossRef] [PubMed]
154. Paniagua, C.; Perry, L.; Benitez-Alfonso, Y. A Phylogenetic and Transcriptomic Study of the β-1,3-Glucanase Family in Tomato Identifies Candidate Targets for Fruit Improvement. *bioRxiv*, 2021; *preprint*. [CrossRef]
155. Beffa, R.S.; Hofer, R.M.; Thomas, M.; Meins, F. Decreased Susceptibility to Viral Disease of [Beta]-1,3-Glucanase-Deficient Plants Generated by Antisense Transformation. *Plant Cell* **1996**, *8*, 1001–1011. [CrossRef]
156. Iglesias, V.A.; Meins, F. Movement of Plant Viruses Is Delayed in a Beta-1,3-Glucanase-Deficient Mutant Showing a Reduced Plasmodesmatal Size Exclusion Limit and Enhanced Callose Deposition. *Plant J.* **2000**, *21*, 157–166. [CrossRef]

157. Zavaliev, R.; Levy, A.; Gera, A.; Epel, B.L. Subcellular Dynamics and Role of Arabidopsis β-1,3-Glucanases in Cell-to-Cell Movement of Tobamoviruses. *Mol. Plant-Microbe Interact. MPMI* **2013**, *26*, 1016–1030. [CrossRef]
158. Li, Z.; Variz, H.; Chen, Y.; Liu, S.-L.; Aung, K. Plasmodesmata-Dependent Intercellular Movement of Bacterial Effectors. *Front. Plant Sci.* **2021**, *12*, 464. [CrossRef] [PubMed]
159. Shi, F.; Wang, Y.; Zhang, F.; Yuan, X.; Chen, H.; Chen, X.; Chen, X.; Cui, X. Soybean Endo-1,3-Beta-Glucanase (GmGLU) Interaction with *Soybean mosaic Virus*-Encoded P3 Protein May Contribute to the Intercelluar Movement. *Front. Genet.* **2020**, *11*, 536771. [CrossRef] [PubMed]
160. Chowdhury, R.N.; Lasky, D.; Karki, H.; Zhang, Z.; Goyer, A.; Halterman, D.; Rakotondrafara, A.M. HCPro Suppression of Callose Deposition Contributes to Strain-Specific Resistance Against Potato Virus Y. *Phytopathology* **2020**, *110*, 164–173. [CrossRef] [PubMed]
161. Alazem, M.; Tseng, K.-C.; Chang, W.-C.; Seo, J.-K.; Kim, K.-H. Elements Involved in the Rsv3-Mediated Extreme Resistance against an Avirulent Strain of Soybean Mosaic Virus. *Viruses* **2018**, *10*, 581. [CrossRef]
162. Otulak-Kozieł, K.; Kozieł, E.; Lockhart, B.E.L. Plant Cell Wall Dynamics in Compatible and Incompatible Potato Response to Infection Caused by *Potato Virus Y* ($PVY^{NTN}$). *Int. J. Mol. Sci.* **2018**, *19*, 862. [CrossRef]
163. Wang, Y.; Li, X.; Fan, B.; Zhu, C.; Chen, Z. Regulation and Function of Defense-Related Callose Deposition in Plants. *Int. J. Mol. Sci.* **2021**, *22*, 2393. [CrossRef]
164. Griffith, M.; Yaish, M.W.F. Antifreeze Proteins in Overwintering Plants: A Tale of Two Activities. *Trends Plant Sci.* **2004**, *9*, 399–405. [CrossRef]
165. Yaish, M.W.F.; Doxey, A.C.; McConkey, B.J.; Moffatt, B.A.; Griffith, M. Cold-Active Winter Rye Glucanases with Ice-Binding Capacity. *Plant Physiol.* **2006**, *141*, 1459–1472. [CrossRef]
166. Romero, I.; Fernandez-Caballero, C.; Goñi, O.; Escribano, M.I.; Merodio, C.; Sanchez-Ballesta, M.T. Functionality of a Class I Beta-1,3-Glucanase from Skin of Table Grapes Berries. *Plant Sci.* **2008**, *174*, 641–648. [CrossRef]
167. Ding, C.-K.; Wang, C.; Gross, K.C.; Smith, D.L. Jasmonate and Salicylate Induce the Expression of Pathogenesis-Related-Protein Genes and Increase Resistance to Chilling Injury in Tomato Fruit. *Planta* **2002**, *214*, 895–901. [CrossRef]
168. Sanchez-Ballesta, M.T.; Gosalbes, M.J.; Rodrigo, M.J.; Granell, A.; Zacarias, L.; Lafuente, M.T. Characterization of a β-1,3-Glucanase from Citrus Fruit as Related to Chilling-Induced Injury and Ethylene Production. *Postharvest Biol. Technol.* **2006**, *40*, 133–140. [CrossRef]
169. Fincher, G.B. Exploring the Evolution of (1,3;1,4)-β-d-Glucans in Plant Cell Walls: Comparative Genomics Can Help! *Curr. Opin. Plant Biol.* **2009**, *12*, 140–147. [CrossRef] [PubMed]
170. Fincher, G.B.; Lock, P.A.; Morgan, M.M.; Lingelbach, K.; Wettenhall, R.E.; Mercer, J.F.; Brandt, A.; Thomsen, K.K. Primary Structure of the (1→3,1→4)-Beta-D-Glucan 4-Glucohydrolase from Barley Aleurone. *Proc. Natl. Acad. Sci. USA* **1986**, *83*, 2081–2085. [CrossRef] [PubMed]
171. Simmons, C.R.; Litts, J.C.; Huang, N.; Rodriguez, R.L. Structure of a Rice β-Glucanase Gene Regulated by Ethylene, Cytokinin, Wounding, Salicylic Acid and Fungal Elicitors. *Plant Mol. Biol.* **1992**, *18*, 33–45. [CrossRef]
172. Lai, D.M.; Høj, P.B.; Fincher, G.B. Purification and Characterization of (1→3,1→4)-Beta-Glucan Endohydrolases from Germinated Wheat (*Triticum aestivum*). *Plant Mol. Biol.* **1993**, *22*, 847–859. [CrossRef]
173. Yun, S.J.; Martin, D.J.; Gengenbach, B.G.; Rines, H.W.; Somers, D.A. Sequence of a (1-3,1-4)-Beta-Glucanase CDNA from Oat. *Plant Physiol.* **1993**, *103*, 295–296. [CrossRef]
174. Akiyama, T.; Jin, S.; Yoshida, M.; Hoshino, T.; Opassiri, R.; Ketudat Cairns, J.R. Expression of an Endo-(1,3;1,4)-β-Glucanase in Response to Wounding, Methyl Jasmonate, Abscisic Acid and Ethephon in Rice Seedlings. *J. Plant Physiol.* **2009**, *166*, 1814–1825. [CrossRef]
175. Kraemer, F.J.; Lunde, C.; Koch, M.; Kuhn, B.M.; Ruehl, C.; Brown, P.J.; Hoffmann, P.; Göhre, V.; Hake, S.; Pauly, M.; et al. A Mixed-Linkage (1,3;1,4)-β-D-Glucan Specific Hydrolase Mediates Dark-Triggered Degradation of This Plant Cell Wall Polysaccharide. *Plant Physiol.* **2021**, *185*, 1559–1573. [CrossRef]
176. Fan, M.; Jensen, J.K.; Zemelis-Durfee, S.; Kim, S.-J.; Chan, J.-Y.; Beaudry, C.M.; Brandizzi, F.; Wilkerson, C.G. Disruption of *Brachypodium* Lichenase Alters Metabolism of Mixed-Linkage Glucan and Starch. *Plant J.* **2022**, *109*, 927–939. [CrossRef]
177. Litts, J.C.; Simmons, C.R.; Karrer, E.E.; Huang, N.; Rodriguez, R.L. The Isolation and Characterization of a Barley 1,3-1,4-β-Glucanase Gene. *Eur. J. Biochem.* **1990**, *194*, 831–838. [CrossRef]
178. Slakeski, N.; Fincher, G.B. Developmental Regulation of (1→3, 1→4)-β-Glucanase Gene Expression in Barley. *Plant Physiol.* **1992**, *99*, 1226–1231. [CrossRef] [PubMed]
179. Slakeski, N.; Baulcombe, D.C.; Devos, K.M.; Ahluwalia, B.; Doan, D.N.; Fincher, G.B. Structure and Tissue-Specific Regulation of Genes Encoding Barley (1→3,1→4)-β-Glucan Endohydrolases. *Mol. Gen. Genet. MGG* **1990**, *224*, 437–449. [CrossRef] [PubMed]
180. Collins, H.M.; Betts, N.S.; Dockter, C.; Berkowitz, O.; Braumann, I.; Cuesta-Seijo, J.A.; Skadhauge, B.; Whelan, J.; Bulone, V.; Fincher, G.B. Genes That Mediate Starch Metabolism in Developing and Germinated Barley Grain. *Front. Plant Sci.* **2021**, *12*, 208. [CrossRef] [PubMed]
181. Fincher, G.B. Molecular and Cellular Biology Associated with Endosperm Mobilization in Germinating Cereal Grains. *Annu. Rev. Plant Biol.* **1989**, *40*, 305–346. [CrossRef]
182. Carpita, N.C.; Defernez, M.; Findlay, K.; Wells, B.; Shoue, D.A.; Catchpole, G.; Wilson, R.H.; McCann, M.C. Cell Wall Architecture of the Elongating Maize Coleoptile. *Plant Physiol.* **2001**, *127*, 551–565. [CrossRef]

183. Kiemle, S.N.; Zhang, X.; Esker, A.R.; Toriz, G.; Gatenholm, P.; Cosgrove, D.J. Role of (1,3)(1,4)-β-Glucan in Cell Walls: Interaction with Cellulose. *Biomacromolecules* **2014**, *15*, 1727–1736. [CrossRef]
184. Smith-Moritz, A.M.; Hao, Z.; Fernández-Niño, S.G.; Fangel, J.U.; Verhertbruggen, Y.; Holman, H.-Y.N.; Willats, W.G.T.; Ronald, P.C.; Scheller, H.V.; Heazlewood, J.L.; et al. Structural Characterization of a Mixed-Linkage Glucan Deficient Mutant Reveals Alteration in Cellulose Microfibril Orientation in Rice Coleoptile Mesophyll Cell Walls. *Front. Plant Sci.* **2015**, *6*, 628. [CrossRef]
185. Carpita, N.C.; Gibeaut, D.M. Structural Models of Primary Cell Walls in Flowering Plants: Consistency of Molecular Structure with the Physical Properties of the Walls during Growth. *Plant J.* **1993**, *3*, 1–30. [CrossRef]
186. Cosgrove, D.J. Enzymes and Other Agents That Enhance Cell Wall Extensibility. *Annu. Rev. Plant Physiol. Plant Mol. Biol.* **1999**, *50*, 391–417. [CrossRef]
187. Roulin, S.; Buchala, A.J.; Fincher, G.B. Induction of (1→3,1→4)-β-D-Glucan Hydrolases in Leaves of Dark-Incubated Barley Seedlings. *Planta* **2002**, *215*, 51–59. [CrossRef]
188. Nishizawa, Y.; Saruta, M.; Nakazono, K.; Nishio, Z.; Soma, M.; Yoshida, T.; Nakajima, E.; Hibi, T. Characterization of Transgenic Rice Plants Over-Expressing the Stress-Inducible β-Glucanase Gene Gns1. *Plant Mol. Biol.* **2003**, *51*, 143–152. [CrossRef] [PubMed]
189. Barghahn, S.; Arnal, G.; Jain, N.; Petutschnig, E.; Brumer, H.; Lipka, V. Mixed Linkage β-1,3/1,4-Glucan Oligosaccharides Induce Defense Responses in *Hordeum vulgare* and *Arabidopsis thaliana*. *Front. Plant Sci.* **2021**, *12*, 1201. [CrossRef] [PubMed]
190. Rebaque, D.; del Hierro, I.; López, G.; Bacete, L.; Vilaplana, F.; Dallabernardina, P.; Pfrengle, F.; Jordá, L.; Sánchez-Vallet, A.; Pérez, R.; et al. Cell Wall-Derived Mixed-Linked β-1,3/1,4-Glucans Trigger Immune Responses and Disease Resistance in Plants. *Plant J.* **2021**, *106*, 601–615. [CrossRef] [PubMed]
191. Pettolino, F.; Sasaki, I.; Turbic, A.; Wilson, S.M.; Bacic, A.; Hrmova, M.; Fincher, G.B. Hyphal Cell Walls from the Plant Pathogen *Rhynchosporium secalis* Contain (1,3/1,6)-β-D-Glucans, Galacto- and Rhamnomannans, (1,3;1,4)-β-D-Glucans and Chitin. *FEBS J.* **2009**, *276*, 3698–3709. [CrossRef] [PubMed]
192. Pérez-Mendoza, D.; Rodríguez-Carvajal, M.Á.; Romero-Jiménez, L.; de Araujo Farias, G.; Lloret, J.; Gallegos, M.T.; Sanjuán, J. Novel Mixed-Linkage β-Glucan Activated by c-Di-GMP in *Sinorhizobium meliloti*. *Proc. Natl. Acad. Sci. USA* **2015**, *112*, E757–E765. [CrossRef]
193. Samar, D.; Kieler, J.B.; Klutts, J.S. Identification and Deletion of Tft1, a Predicted Glycosyltransferase Necessary for Cell Wall β-1,3;1,4-Glucan Synthesis in *Aspergillus fumigatus*. *PLoS ONE* **2015**, *10*, e0117336. [CrossRef]
194. Ao, J.; Free, S.J. Genetic and Biochemical Characterization of the GH72 Family of Cell Wall Transglycosylases in *Neurospora crassa*. *Fungal Genet. Biol.* **2017**, *101*, 46–54. [CrossRef]
195. Woodward, J.R.; Fincher, G.B. Substrate Specificities and Kinetic Properties of Two (1→3),(1→4)-β-d-Glucan Endo-Hydrolases from Germinating Barley (*Hordeum vulgare*). *Carbohydr. Res.* **1982**, *106*, 111–122. [CrossRef]

MDPI
St. Alban-Anlage 66
4052 Basel
Switzerland
Tel. +41 61 683 77 34
Fax +41 61 302 89 18
www.mdpi.com

*Plants* Editorial Office
E-mail: plants@mdpi.com
www.mdpi.com/journal/plants

www.ingramcontent.com/pod-product-compliance
Lightning Source LLC
Chambersburg PA
CBHW062104220926
51850CB00010B/2880